The Ludic City

CW01501050

Why do we have public spaces in cities? What are they for? What role do they have in everyday social life? *The Ludic City* argues that one of the fundamental functions of public space is as a setting for informal, non-instrumental social interaction, or play. The concept of play highlights the distinctive character of urban experience: the ways people sense urban settings, move through them and act within them. Play is an important but largely neglected aspect of people's experience of urban society, and embraces a wide variety of activities which are spontaneous, irrational or risky, and which are often unanticipated by designers, managers and other users. Focusing on the playful uses of public space, this book provides a much-needed counterpoint to the instrumental pragmatism which dominates everyday urban life and the design of city spaces.

Drawing together arguments from the fields of urban design, planning, sociology, anthropology, philosophy and environmental psychology, this book provides a fresh and detailed depiction of play in the specific context of urban public space. By illustrating the forms that play takes, it reveals people's creativity, curiosity and imagination in using urban space.

The Ludic City draws upon extensive observation of behaviors in public spaces, with detailed studies of Melbourne, London, Berlin, New York and Brisbane. The findings suggest that even the most typical street corner or innocuous doorway can be a site for risk-taking, the display of identity and testing the limits of one's own abilities.

The book will provide urban designers, policy-makers, planners and researchers with an awareness of how a playful, non-reductive understanding of space and social practice can positively shape urban design practice and public policy.

Quentin Stevens is Lecturer in Planning and Urban Design at the Bartlett School of Planning, University College London. He has a PhD in Urban Design from the University of Melbourne and has studied Architecture and Urban Planning in Melbourne, Berkeley and Chicago. He has also worked as an urban designer and planner in both Australia and the United States and is co-editor of *Loose Space: Possibility and Diversity in Urban Life*.

The Ludic City

Exploring the potential of public spaces

Quentin Stevens

 Routledge
Taylor & Francis Group

LONDON AND NEW YORK

First published 2007
by Routledge
2 Park Square, Milton Park, Abingdon, Oxon OX14 4RN

Simultaneously published in the USA and Canada
by Routledge
270 Madison Avenue, New York, NY 10016

Routledge is an imprint of the Taylor & Francis Group, an informa business

© 2007 Quentin Stevens

Typeset in Sabon by
Florence Production Ltd, Stoodleigh, Devon
Printed and bound in Great Britain by
The Cromwell Press, Trowbridge, Wiltshire

British Library Cataloguing in Publication Data
A catalogue record for this book is available from the British Library

Library of Congress Cataloging in Publication Data
Stevens, Quentin, 1969–
The Ludic city: exploring the potential of public spaces /
Quentin Stevens.
    p. cm.
  Includes bibliographical references.
1. Sociology, Urban. 2. Public spaces. 3. Play. I. Title.
HT153.S833 2007
307.76–dc22   2006034797

ISBN10: 0–415–40179–8 (hbk)
ISBN10: 0–415–40180–1 (pbk)

ISBN13: 978–0–415–40179–1 (hbk)
ISBN13: 978–0–415–40180–7 (pbk)

The city must be a place of waste, for one wastes space and time; everything mustn't be foreseen and functional . . . the most beautiful cities were those where festivals were not planned in advance, but there was a space where they could unfold.

(Lefebvre 1987: 36)

Contents

Acknowledgements

Kim Dovey's contribution to this work has been tremendous. The same goes for my academic development. Kim started teaching me right from my first week of university. Twenty years later, I have a lot to thank him for, and I hope we continue to have heated discussions for many years to come. Special thanks also to Karen Franck for her constant support, critique and insight. Thank you to George Hemmens for introducing me to the ideas of Henri Lefebvre and to the essays in Sorkin's *Variations on a Theme Park*, and to Dana Buntrock for haranguing me constantly until I finally realized I would never really be satisfied until I did a PhD. You were right. The following people also deserve mention for their useful comments, advice and assistance: Iain Borden, Matthew Carmona, Hans-Peter Dreitzel, June Factor, John Friedmann, Bernardo Jiminez, Sandra Kadji-O'Grady, Ross King, Loretta Lees, Marissa Lindquist, John Macarthur, Kevin McDonald, Donald McNeill, Filipa Matos, Greg Missingham, David Naegler, Daniel Ross, Leonie Sandercock, Louie Sieh, Mark Tewdwr-Jones, Fran Tonkiss, Anna Tweeddale, Stephen Wood and Ian Woodcock. Thanks also to artist Candy Stevens for her excellent work on the book's cover. Lastly I wish to thank my parents, as well as Simone, Victoria, Sally and Dagmar.

Some of the ideas in this book have been explored previously in other publications: 'The Shape of Urban Experience: A Re-evaluation of Lynch's Five Elements', *Environment and Planning B*, 33(6), 2006; 'Situationist City' in *The International Encyclopaedia of Human Geography*, Oxford: Elsevier, 2007; 'Betwixt and Between: Building Thresholds, Liminality and Public Space' in K. Franck and Q. Stevens (eds) *Loose Space: Possibility and Diversity in Urban Life*, London: Routledge, 2006; 'The Design of Urban Waterfronts: A Critique of Two Australian "Southbanks"', *Town Planning Review*, 77(2), 2006; 'Appropriating the Spectacle: Play and Politics in a Leisure Landscape' (with Kim Dovey), *Journal of Urban Design*, 9(3), 2004; 'Urban Escapades: Play in Melbourne's Public Spaces' in L. Lees (ed.) *The Emancipatory City? Paradoxes and Possibilities*, London: Sage, 2004; 'Espectaculo y Carnaval: la dialectica del esparcimiento en espacios publicos' ('Spectacle and Carnival: The Dialecticity of Leisure in Public Spaces'), *PISO*,

City at Ground Level, Guadalajara, Mexico, 1(1), 2003; 'The Dialectics of Urban Play' in R. G. Mira, J. M. S. Cameselle and J. R. Martinez (eds) *Culture, Environmental Action and Sustainability*, Göttingen: Hogrefe and Huber, 2003; 'A Fragile Resource', *Brisbane Courier Mail*, 12 November 2002.

All maps and photographs are copyright of the author.

Introduction

What are public spaces for? Urban design often pursues such clear-cut instrumental goals as comfort, practicality and order. But the scope of everyday life in urban spaces is never completely subordinated to the achievement of predefined, rational objectives. People can be capricious and unpredictable. Urban spaces and the activities which occur in them constantly generate disorder, spontaneity, risk and change. Urban public spaces offer a richness of experiences and possibilities for action.

This book explores the playful uses of urban spaces. Play is an important but largely neglected aspect of people's experience of urban society and urban space. It involves controversial expenditures of time and energy, 'unfunctional', economically inefficient, impractical and socially unredemptive activities which are often unanticipated by designers, managers and other users. Play reveals the potentials that public spaces offer.

The ways that people's needs are expressed through their playful behavior in urban public space both extend beyond simple definitions of function and run contrary to the idea of function. The density and diversity of city life inevitably leads to tensions and contradictions between rational social organization and people's other desires. Non-instrumental, playful behavior thrives on a continuing negotiation with the various forms of discipline, exploitation and spectacle which constitute the contemporary city. Play concentrates attention on practices which have a dialectical relation to the order, fixity and functional and semiotic determinism of built form.

To date there has been precious little focused empirical study of urban public spaces which can illuminate a 'non-functional' understanding of the use and design of public space: Lennard and Lennard (1984), Dargan and Zeitlin (1990) and Borden (2001b) are the few relevant works. Whyte (1980, 1988) and Gehl (1987) study where people choose to spend time in public spaces, and then offer clear and detailed analysis of the features of those environments that make people want to stop there. They thus relate non-instrumental behavior to urban design. However, their interest is mostly space-centered, looking at the general categories of everyday behavior of people in public spaces, the great majority of which are reasonably mundane, pragmatic and predictable. They illustrate much about how and why public spaces work and do not work, but they only hint at the full variety of unexpected and impractical uses which people make of public spaces.

This book draws together observations of a wide range of everyday play activities in public spaces in a number of large cities – Melbourne, London, Berlin, New York and Brisbane – since the mid 1990s. What counts as 'playful' behavior clearly varies between and within cultures. The diversity, leisure and *jouissance* generated by urban life are also constantly being reappropriated and exploited by governments and investors to serve instrumental ends of power and profit. What appears to be play is by no means universal, nor is it always free and benign. Play is contingent, it exists among the tensions and contradictions of urban social life. It is first and foremost the mechanics of behavior, and not the place or social context, which are of interest here. Focusing on the less practical forms of social action sheds new light on the role that urban design plays in action. The kinds of social practices which can be labeled as 'play' are a fresh lens for viewing urban settings and for understanding their performance.

The idea that public space is not merely instrumental presents a challenge to those responsible for designing and managing urban space. The discipline of urban design has at its heart a very vague, abstract and potentially ambiguous concept: amenity. The concept of amenity lies at the nexus of two different fundamental issues. The first is the philosophical question of what makes a good environment, the desired mix of potentials and challenges which a setting should provide. Designing the public realm requires calibrating and serving the diverse needs of multiple individuals. Design and management which meet some groups' needs for physical and psychological comfort can place significant constraints upon the desire-fulfillment of others, and 'if public spaces prioritise one kind of need, then people not motivated by that need will be inclined to stay away' (Mean and Tims 2005: 52). It is important to understand all the uses of the city, however unconventional, because the openness and publicness of urban space gain their meaning through the breadth of users and the varieties of actions that are pursued there. The second issue is the question of how spatial characteristics shape people's experiences and behaviors. Amenity thus presumes an idea of function, in terms of both what resultant actions or experiences are desirable, and the ways physical environments help to make those outcomes possible. However, neither people's desires nor their actions are sovereign, well-understood and fixed; both are contingent, a product of circumstance, and continually changing. It is difficult to predict the timing or location of unpragmatic activities like play. They tend to confound expectations. The amenity of the public realm thus means more than serving predetermined, practical functions. It can also be about the potential which design provides for expanding people's experience and their capacities.

The first three chapters that follow provide a theoretical base for defining and understanding play in public space, working from general theories of urban society down to the specific dimensions of the human body. Chapter 1 is a broad critique of urban social life, following a theoretical lineage which stems from Surrealism. This discussion is organized around the idea that people's experiences in urban space are characterized by a range of tensions: tensions between exchange value and use value, between the needs and actions of the collective and the individual, between alienation and participation, between the instrumental rationality of work and the creative freedom of play. The city is the most intense manifestation of the tensions of modern life, but urban public spaces also provide the conditions for resolving these tensions, by stimulating playful practices. Chapter 2 provides a definition of urban play behavior which draws primarily on anthropological and sociological knowledge. This framework identifies a range of oppositions between play and serious, productive activity, including the kinds of spatial and temporal conditions which set play apart from the everyday. Four distinct types of play are defined, emphasizing different ways that play transgresses norms of bodily action and perception and the body's relation to space. Encounters with strangers define the public dimension of play. Chapter 3 examines the physical dimensions of sensory perception and social engagement which frame playful possibilities, focusing on bodily relations among strangers in public settings.

Chapters 4 to 8 provide detailed descriptions of play behavior in a diversity of urban public spaces. The chapters are organized around five kinds of settings where most play activities seem to occur: paths, intersections, boundaries, thresholds and props. Various playful acts are made possible by the formal qualities of these types of spaces. The everyday activities commonly associated with these spaces also help stimulate the possibility of play. These chapters examine how each setting relates to the main aspects of play covered in the theoretical chapters which precede them: the bodily mechanics of play actions, the spatial relations between bodies and the scope of sensory perceptions which surround people who play.

The concluding chapter is a consideration of how play might help to shape an agenda for the design and management of urban public spaces and the planning of urban areas. It explores three themes that are crucial to the wider experiential potential of urban space: functionality, access and public performance. These themes focus attention on the ways in which urban design can serve the definitions on play mapped out in Chapter 2, specifically in relation to the concepts of non-instrumentality, spatial separation and publicness. A fourth overarching theme explored is how different elements of urban structure and their arrangement frame particular experiential possibilities, and thereby specific forms of play.

The theoretical perspectives on play, instrumentality, chance, distraction and risk outlined in the early chapters, and the observational findings which follow, provide ample grounds to reconsider urban design's core knowledge base regarding the function, amenity and perception of urban space – the work of pioneers such as Lynch, Jacobs, Alexander and Gehl – and to rethink such venerable concepts as The Image of the City; the need for mixed primary uses and buildings of different ages; the distinction between necessary and optional activities; the uses of sidewalks and the need for small blocks; the flexibility of particular urban forms; and the ways in which urban activities, experiences and spaces are related to each other. Existing knowledge about the relations between built form and behavior turns out to also have a certain amount of unexpected usefulness for understanding more complex and unanticipated uses of open space.

Yet play by its nature remains creative, unpredictable and hard to qualify. Rather than trying to provide a definitive answer to the question of what urban public spaces are for, or should be for, the argument put forward in this book is that design and management need to ensure that the question remains open, to allow a certain amount of play. People are all different, and they play differently. No book can pretend to account for the full scope of ways in which people enjoy the freedoms of urban open space. The observations here do, however, illustrate some of the limits of urban design thinking and practice, as well as people's great capacity to look beyond those limits and discover surprising potential in urban spaces.

Chapter 1

Urban conditions and everyday life

Cities are typically seen as the engines of modern economic life. Cities are thus principally planned to optimize work and other practical, rational, preconceived objectives, and are designed accordingly, with even leisure space serving well-defined functions. But people do not only gather together in cities to meet their basic physiological needs; they also come to cities searching for love, esteem and self-actualization, and to experience the diversity of the world around them and to learn to understand it (Maslow 1943). Cities have a wide range of functions and they serve a wide range of aspirations (Mumford 1961, 1996). Wirth famously defined the urban condition as 'a relatively large, dense and permanent settlement of socially heterogeneous individuals' (Wirth 1996: 190–91). But this expression misses the heart of the matter: it is the interactions among these diverse individuals, their mixing, which really constitutes urbanity, and which gives city life its special character and possibility. Urbanism without a certain degree of cosmopolitanism is just a mass of completely unconnected, alienated strangers. It is in public open spaces that people are best able and most likely to engage with the social diversity gathered together in cities.

The complexity and breadth of reasons that people are attracted to cities becomes more obvious when focus is shifted from the outcomes of their actions to the qualities of the experience itself. In this book, analysis of the special character of life in urban public space, as a milieu for people's experiences and actions, focuses on two main elements: the particular dynamics of urban social relations, and a phenomenological account of urban spaces as perceived by people who use them.

The exploration of this theme is guided primarily by the writings of Henri Lefebvre, Walter Benjamin and the Situationist International. These authors all develop definitions of urbanism which they link 'upward' to a wider critique of contemporary society, and also 'downward' to an analysis of the everyday behavior of individuals in urban space. Their understandings of urbanism also present a conscious critique of the mainstream philosophies that purport to explain urban life.

Both Wirth's 'ecological' model of urban society and its Marxian critique through political economy share two major limitations in their characterization of urban life (Gottdiener 1985). First, they cannot explain voluntaristic actions outside their own conceptual frameworks: on the one hand, instrumentally rational choice and on the other, the social relations of production. Second, they deny that the specific perceptions and behaviors of people within material space have any real significance for their social existence; indeed, they conceive of social relations as existing independently of space. Public space and the events that happen there are epiphenomena of society, or, at worst, a delusion, a false consciousness.

Lefebvre (1991b) argues that all forms of social experience are constituted in and through space. It is in urban spaces that the scope of what people experience as 'everyday life' continually develops. Lefebvre identifies three distinct aspects of the experience of urban space as a social milieu, his well-known 'conceptual triad'. *Spatial practices* include all the material social interactions occurring within space to produce and reproduce a particular social formation. Put simply, these practices are what 'actually' occurs; they have direct physical and social consequences. *Representations of space* are the social codes through which people discuss and understand material space and spatial practices. These conventions include names and descriptions of places. People's perceptions of the 'reality' of social life in space are filtered through what they 'understand', or 'believe', and how they come to know it. *Representational spaces* are spaces as lived by their inhabitants through complex symbolic association and imagery; 'it overlays physical space, making symbolic use of its objects' (Lefebvre 1991b: 39). These imaginings include real locations such as symbolic sites, as well as mental inventions of new possibilities for spatial practices: ideal, dystopian or alternative social and spatial formations. Representational spaces, whether they have material existence or not, are meanings, references to ideas about social space.

Lefebvre notes that these three aspects are always present and interrelated in any space which might be analyzed. Most critics of urban space examine spatial practices; Lefebvre's concepts of *representations of space* and *representational spaces* extend beyond materialist analyses and emphasize the importance of the perception and constitution of meaning to the definition of what space is. However, if the aim is to understand the everyday social reality of the city, examining representations of space presents problems. Representations of space 'tend ... toward a system of verbal ... signs' (Lefebvre 1991b: 39): they are designed as a form of knowledge, distanced and abstracted from experience. They are shaped to be communicated – with clear purposes in mind – and not to be lived directly, and hence they suppress many of the contingent nuances of practices and inventions. They are also shaped as part of the broader power relations operating in society, and hence they tend to be more sharply ideological than the other aspects of space. Space as conceptualized 'is the dominant space in any society ... tied to the

relations of production and to the "order" which those relations impose'
(Lefebvre 1991b: 38–39, 33). Such terms as 'urban', 'public' and 'open space'
both inscribe and disguise social power relations; they do not necessarily
bring closer knowledge of what life in urban space is really like.

Urban experience and social needs are more than mere conceptual
abstractions; they can be understood by looking at everyday life on the
streets, at its specific and diverse qualities, at the meanings it might have for
those who live it, and in particular at the complex tensions which arise
between different needs, different meanings and different users in spaces.
Lefebvre, Benjamin and the Situationists also all identify social practices of
play as key to understanding the dynamic tensions which shape everyday
life in public. This chapter explores in turn the way that urban conditions
shape particular dynamics of spatial behavior, sensory perception, and needs,
closing with an examination of the role of leisure within everyday urban
life. The idea of play constantly emerges within this discussion as a key aspect
of urban experience, although different facets of its significance are revealed.

Urban spatial practices

Both Chicago School and structuralist Marxist thinking on the city focus
around the idea that the organization of urban spatial patterns and practices
are largely determined by economic forces. These forces aim to optimize
production and consumption by different social groups through economies
of scale and agglomeration or opportunities for the expropriation of profit.
Lefebvre contributes to this argument an analysis of how capital markets
and the state organize urban space to produce the social relations of industrial
mass production (Gottdiener 1985). However, 'There is not . . . a strict
correspondence between modes of production and the spaces they constitute'
(Lefebvre 1987: 31).

At the base of Lefebvre's own theorization is an analysis of urban space
and urban life as a social fact. He argues that urbanism is not merely an
induced effect of rational production. His proof is historical: cities existed
in pre-capitalist times and have served a broad range of social functions
in addition to enhancing production. Each city can be understood as a
comprehensive, distinctive cultural artifact and a complex totality of cultural
practices both old and new. These different facets are embraced by his use
of the term *oeuvre* (Lefebvre 1996). His critique of 'functional' urban
planning and his alternative vision can be summarized thus:

> [The city] figures in planning as a cog: it becomes the material device
> apt to organize production, control the daily life of producers and the
> consumption of products . . . It did not have, it has no meaning but as
> an *oeuvre*, as an end, as a place of free enjoyment, as domain of use value.
>
> (Lefebvre 1996: 126)

Lefebvre goes on to identify two essential ways in which cultural life and social life struggle to find their realization through urban spatial conditions (Martins 1982). The first of these is the assembly of the full diversity of the population and their activities, their focused centrality in space and simultaneity in time, and their participation in the management and development of social space:

> The *right to the city* . . . stipulates the right to meeting and gathering; places and objects must answer to certain 'needs' generally misunderstood, to certain despised and moreover transfunctional 'functions': the 'need' for social life and a centre, the need and the function of play, the symbolic function of space.
>
> (Lefebvre 1996: 195, emphasis in original)

Urban culture is not a tidy, static fact, and its practices will inevitably be multiple, contradictory and dynamic. Economies of scale in the city support specialization, which leads to further differentiation of social identities and occupations. Specialization diminishes the significance of the extended family as the locus of social life, and leads to more numerous and complex interrelations between heterogeneous individuals who have no primary ties (Wirth 1996). The processes of growth and differentiation in cities mean there is a certain amount of instability and change in urban living.

Second, 'The Right to Difference' means 'not being classified within pre-established categories' (Martins 1982: 183). While industrial mass production attempts to homogenize urban spatial activity, urban society differentiates (Lefebvre 1991b). Because of its diversity, urban social life brings about the satisfaction of a wide range of human needs. Yet it also modifies and creates new needs, and people constantly struggle to reshape social space to reflect and to serve these new needs (Lefebvre 1991a, 1991b).

These two themes of concentration and diversity are familiar from Wirth's formulation of the urban. Lefebvre also highlights a third distinct condition, which relates to the other two, but is something new: the city frames opportunities for play (Lefebvre 1991b, 1996). While a variety of instrumental imperatives may cause people to live together in cities, dense spaces and heterogeneous populations can make a significant contribution to social development only where there are chance encounters, social mixing, exploration of the unfamiliar and risk; when there is an escape from instrumental social relations. These are the part of the social *oeuvre* which finds its fulfillment in the open public spaces of cities. While most productive work and social reproduction occur in carefully framed settings, play thrives on the density and diversity of people and experiences to be found in urban public space. The concept of play embraces many of the forms of urban social life which can be appreciated as having use value, as ends in themselves. Yet at the same time, play is a lived critique of instrumentally rational action, because it discovers new needs and develops new forms of social life. Lefebvre proposes that it is practices of play which best illustrate the capacities for social action and expression which the urbanization of society has made possible.

The density and diversity of people gathered together in cities give urban social life a distinctive character: it is fundamentally about encounters and interactions among people who are different, and through such interactions the discovery and realization of diverse needs. It is within public spaces that many of these contacts occur. Lefebvre describes the particular social ambience surrounding these urban encounters thus:

> The form of the urban, its supreme reason, namely simultaneity and encounter, cannot disappear ... as a place of encounters, focus of communication and information, the *urban* becomes what it always was: place of desire, permanent disequilibrium, seat of the dissolution of normalities and constraints, the moment of play and of the unpredictable.
>
> (Lefebvre 1996: 129, emphasis in original)

Here Lefebvre makes several distinct points about how and why urban encounters are playful. The city is a site for multiplicitous practices of desire and not only of systematic, instrumental necessity. Many social encounters in cities are sudden, unplanned and unpredictable, and this means urban behavior is spontaneous and creative. Urban life means engaging with and developing behavior which is unfamiliar, testing the usefulness of pre-existing social rules and roles (Phillips and LeGates 1981). Thus it is that urbanization weakens the traditions, conventions, rhythms and social structures which generally guide practical action, broadening the *oeuvre*:

> the development of cities . . . the concentration of different ethnic and/or professional groups in the same space, with in particular the overthrow of spatial and temporal frameworks, favors the confrontation of dif-ferent cultural traditions, which tends to expose their arbitrariness *practically*, through first-hand experience, in the very heart of the routine of the everyday order, of the possibility of doing the same things differently, or, no less important, of doing something different at the same time.
>
> (Bourdieu 1977: 233, emphasis in original)

A wide scope of activities, both practical and playful, occur in any given urban space at different times, under different conditions, and often even at the same time or under the same conditions. This dynamic social context affects the way individual acts are conducted. Bourdieu uses a spatial illustration to explain his concept *habitus* which helps illustrate the impact of urban space on the conduct of those who use it. He refers to the different social positions and social distances of people brought together in physical space, depicting *habitus* as 'so many reminders of this distance and of the conduct required in order to "keep one's distance" or to manipulate it strategically' (Bourdieu 1977: 82). The proximity and chance nature of encounters in urban public space frequently disturb such expectations, and heighten tensions. As Edensor (1998: 217) argues, 'disorganized' urban spaces 'challenge the physical and mental dispositions . . . by confrontation with different orders of sensory experience, social interaction, regulation and movement'. Sennett (1974: 49) similarly notes that public space is an 'amorphous milieu . . . where no-one is really sure what appropriate standards of behavior are', and that the strangers encountered in the city are 'unknown quantities'. Nevertheless, many of the everyday actions people perform in the city are significantly shaped by their publicness, the extent to which they involve or are consciously directed toward strangers (Lofland 1998). The acknowledgement and mediation of the mutual impacts of people's actions leads to the development of social graces, 'behavior which all agree to treat arbitrarily as "proper" and believable'. This gives people 'means to be sociable, on impersonal grounds' (Sennett 1974: 49, 64). People must adapt their own perceptions, inclinations and abilities to suit the unexpected, unfamiliar circumstances of urban social space. And yet in the majority of the many potentially tense social situations which arise in daily urban life, people ignore or tolerate the strange behavior of others, and it is this 'sense of freedom from judgment that many people report . . . as a major pleasure of being "out in public"' (Lofland 1998: 32).

The processes of growth and differentiation in cities also mean there is a certain amount of instability and change in the individual's own way of life (Sennett 1971). Mumford portrays the city as an engine of continuous social development:

> In the city, the making and remaking of selves . . . is one of its principal functions . . . each urban period provides a multitude of new roles and an equal diversity of new potentialities. These bring about corresponding changes in laws, manners, moral evaluations, costume, and architecture, and finally they transform the city as a living whole.
>
> (Mumford 1996: 116)

The city provides the opportunity for people to address their various needs by exploring new possibilities in life, by expanding their *habitus*, in large part through their encounters with others. Sennett (1974) points out that the accumulation of surplus wealth in the city engenders a leisure lifestyle, where social relations are released from the burden of functional necessity. In such circumstances, one is likely to encounter strangers in the city in situations where 'you are not meeting for some functional purpose, but meeting in the context of nonfunctional socializing, of social interaction for its own sake' (Sennett 1974: 118). The playful theatricality of roles and masks in urban encounter can thus be understood as an expression of 'natural passions' which '[transcend] work, family and civic duty', transgressing and reshaping the rules of social engagement (Sennett 1974: 116). One's sense of self becomes developed through the manipulation of one's appearance in the eyes of strangers. It is primarily in urban public spaces that individuals can act publicly, communicate with the public. Some actions, such as performances of identity and political statements, need to be displayed publicly if they are to have any meaning or purpose at all (Arendt 1958). For Lefebvre play between the various parts of the social whole, unfettered, unpredictable, and above all *expressive* engagement among the full diversity of persons and practices, is a key purpose and outcome of the centralizing function of urban space. This playful function of society can be considered to be just as significant to the constitution of cities as rationality and productivity.

Having presented play as a basic component of use value, as a 'function' of cities, Lefebvre also draws upon play as an example when he looks in detail at the impact of capitalism upon urban space. There are three key features of the transformation of urban space by capitalism. It is homogenized and fragmented, so that it can be exchanged as a commodity, and put in the service of accumulation. This fragmented space is a matrix of 'determinate locations of production and consumption' (Lefebvre 1991b: 341) which 'divide life into closed, isolated units' (The Lettrist International 1996: 44). At the same time urban space is strategically hierarchized and revalorized (both concretely and symbolically) into centers and peripheries which both 'reflect *and contribute to* the overall social hierarchy' (Martins 1982: 178, emphasis in original). Free-time behavior is also subjected to spatial reorganization under capitalism, to serve goals of social domination and accumulation:

> witness the predominance of 'amenities', which are a mechanism for the localization and 'punctualization' of activities, including leisure pursuits, sports and games. These are thus concentrated in specially equipped 'spaces' which are as clearly demarcated as factories in the world of work . . . within a space which is determined economically by capital, dominated socially by the bourgeoisie, and ruled politically by the state.
>
> (Lefebvre 1991b: 227)

The Situationists likewise critiqued city planning's creation of 'reservations for "leisure" activities separated from the society', suggesting that '[no] spatiotemporal zone is completely separable' (Kotanyi and Vaneigem 1996: 117). They presented the concept of 'unitary urbanism' as both a critique of and a response to the capitalist city, with leisure providing the best illustration of potential freedom and of refutation:

Even if, during a transitional period, we temporarily accept a rigid division between zones of work and residence, we should at least envisage a third sphere: that of life itself (the sphere of freedom, of leisure – the truth of life). Unitary urbanism acknowledges no boundaries; it aims to form a unitary human milieu in which separations such as work/leisure or public/private will finally be dissolved.

(Debord 1996b: 81–82)

Capitalist fragmentation of the city and its social practices has a negative impact on the potentials which urban conditions provide for leisure and play. Social practices of play are stimulated by the density and diversity of urban populations and their actions, but some aspects of these practices are lost if they become tied to functionally and symbolically determined locations.

A third context within which Lefebvre evokes play is when he describes how the distinct social processes of urbanization and industrialization are linked dialectically. The social relations of urbanism reveal the contradictions of the abstract social relations of production. Lefebvre depicts the central tension in contemporary society's use of urban space as being between the abstract space of production, structured by exchange value and defined and reified by representations of space, and 'social space, or the space of use values produced by the complex interactions of all classes in the pursuit of everyday life': in other words, space as lived (Gottdiener 1985: 127). In terms of social practices,

urban life . . . attempts to foil dominations, by diverting them from their goal . . . In this way the *urban* is more or less the *oeuvre* of its citizens instead of imposing itself upon them as a system, as an already closed book.

(Lefebvre 1996: 117, emphasis in original)

Lefebvre thus also presents play as an important tactic in the struggle over space where use value seeks to elude and overcome the strictures of exchange value and imagination tries to surmount the limits of present realities.

In summary, Lefebvre suggests that individuals' aspirations for urban social practices are enabled by the assembly of social differences and a diversity of activities in space. This concentration brings about conditions of chaos, chance and change, the breaking down of structures and constraints. These conditions encourage and support play as a mode of engaging with difference. The segmentation of social life by capitalism highlights the threat which play poses, as evidence of a non-instrumental, non-commodifiable basis for urban social relations. It also reveals the special potential of play to respond dialectically to instrumentality.

The perception of urban space

The urban social practices which Lefebvre describes are all shaped by the context of actors' perceptions, understandings and expectations about social rules, about 'the public', and about spaces. This perception and interpretation of spaces, practices and meanings is an active process; meanings and rules are not just passively received. The key concept which distinguishes Lefebvre's notion of representational spaces is symbolism. Representational spaces, which in part exist in people's imaginations, tends toward non-verbal symbols and signs. Notions about what the city means and what it communicates are also a key focus in the writings of Walter Benjamin. Many themes in Benjamin's work parallel Lefebvre's thesis, examining the connections between capitalism and specifically urban social relations. Benjamin's interest lies in the perception of the city itself; the ways the city frames perception of other objects and aspects of social life; and playful actions which respond to those perceptions. Benjamin, like Lefebvre, is critical of the kinds of cultural messages which are encoded in urban form.

Savage (1995) describes Benjamin's interest in cities as residing in the relationships between history, experience, memory and the physical environment. The urban built environment is a distinctive symbolic medium. It is encoded and decoded with historical traces and other cultural information in ways which are specific to it, and which differ from the symbolism of other media such as literature and film. People's perceptions of this information subsequently impact the way they behave in the city.

One significant difference between the city and other media is that even in an age of industrial mass production, mass media and mass consumption, cities retain their specificity. This view reinforces the distinction which Lefebvre makes between urbanism and industrialization. Each city has distinctive character which is linked to local landscape, climate and materials. The built form of each city also retains an historical depth of relations to patterns of social behavior (traditions, conventions, techniques), and social meanings that are continually built up through association with these behavior patterns. Cities retain an aura, and urban experience retains connections to this deep and robust reality. This auratic, individual quality is illustrated in Benjamin's descriptions of Naples, Moscow and Marseilles.

The city, as an object of perception, remains at a distance from the observer. It retains an otherness and does not easily yield to the consumption of meaning. Although it connects to memory, it cannot be fully known. It is both familiar and unfamiliar to the gaze. The city can be a strange and terrifying place. There is difficulty and risk involved in engaging with it and trying to understand it. The difficulty of perception comes from people's lack of control over the way they encounter history and memory in the city. Benjamin's writing suggests that urban space and urban social life have a specific impact on how objects (and hence meanings) are presented to the senses and the kinds of relationships in which things are perceived. He argues that the city transforms the character of experience: it intensifies, stultifies, diminishes, fetishizes and sequesters (Gilloch 1996). The nature of the experience of meaning in urban spaces is best evoked in Benjamin's depiction of the city as a labyrinth. The city is a collection of objects shoved together in confusion, without any overriding order or purpose to their communication. Urban perception is characterized by multiplicity and complexity: 'urban spaces are, if anything, "over-inscribed": everything therein resembles a rough draft, jumbled and self-contradictory' (Lefebvre 1991b: 42). For the individual walking through the city, images of the past and present are confronted at random and can be freely associated. These images may provide surprise (even involuntary) triggers for memories of a collective history, traditions and rituals which had been forgotten through subsequent physical and social changes (Buck-Morss 1991). Such images are also an important way that culture is transmitted and reproduced (Lefebvre 1991b).

Both Benjamin and Lefebvre had acquired from the Surrealists the view that urbanism is both something objective and something dreamt or mythological (Buck-Morss 1991). Bourdieu's explanation of the mythological potential of material space is that it becomes structured by 'objectifying operations which the mind applies to it', by 'principles of vision' framed within *habitus* (Bourdieu 1977: 91, 1998: 8). The reality of the world is the sense that people make of it. Inhabited space 'objectifies' the 'mythico-ritual' symbolic structures of the world, and becomes a 'tangible classifying system [which] inculcates and reinforces the taxonomic principles . . . of this culture' (Bourdieu 1977: 89). That is to say, material space helps to frame and reproduce structures of meaning. While Bourdieu suggests the 'symbolic products' of *habitus*, including works of art and myths, have 'an educative effect' which helps to reproduce *habitus* (Bourdieu 1977: 217 fn40), the experience of myths and other symbols in the city can lead to learning which is open-ended and not merely reproductive. Benjamin's arguments suggest that urban space does not entirely make sense. It also often disturbs schemes of perception.

Benjamin and Lefebvre consider two sides of mythology in the modern city. On the one hand, cities are the locus of bourgeois false consciousness: the myths of modernity, including individualism and progress (Gilloch 1996; Lefebvre 1996). The social relations of the modern city are mythologized in part through spatial representations. Framed by the mythic demands of instrumental exchange and mass production and utility, the activities and meanings of life becomes fragmented into separate spaces. One realm is private life, where an individual can supposedly 'be themselves' and look inward to constitute meaning, although it is now primarily mass-produced commodities which organize meaning (Benjamin 1997). Leisure becomes confined to specific, separated spaces such as cafés which reinforce and reproduce bourgeois values and social relations (Lefebvre 1971, 1991a,

1991b). Urban public life, on the other hand, becomes reconstituted as 'an immense accumulation of spectacles' where '[all] that once was directly lived becomes mere representation' (Debord 1994: 12). The city structures new social relations through the collective consumption of mass-produced commodities and experiences, in particular mass media symbolism which provides the false consciousness of social unity, disguising the contradictions at the heart of modern social relations of mass production. In contrast to pre-industrial society, where the specificity of place and action framed and legitimized habitual social practices, the mass production and consumption of commodities and symbols in the modern city engenders instrumental reactions from the individual (Buck-Morss 1991; Gilloch 1996).

Benjamin also considered the modern city to hold the potential for demythification and the creation of new meanings. In common with the Surrealists, he 'viewed the constantly changing new nature of the urban-industrial landscape as itself marvelous and mythic' (Buck-Morss 1991: 256). Unlike other auratic forms of art, which are experienced through concentration, the complexity of urban imagery is perceived in a state of distraction. Reading and interpretation of urban form and spatial practices occurs while people are pursuing other intentions. In contrast to the focus of vision, cities assault all the senses continuously. Cities are not locked into specific modes of receiving or absorbed by the conceptual frames of reference which surround other auratic art forms. The viewer thus also retains a critical distance from the medium:

> Cities, as built environment, contain the potential for the recovery of memory which is an essential element in redemption, yet they avoid the conservative, cultic, ritualistic elements which usually wrap around the auratic object.
>
> (Savage 1995: 212)

While historical meanings are encoded in urban form, perception is also the active process of decoding and employing these meanings in practice. A person's experience of the city triggers memory, but it does not compel them to relate to it in a specific fashion. People's incidental engagements with urban artifacts mean that memory and meaning (and hence, social conventions) are themselves also encountered in a state of distraction, and it becomes possible to recognize, question and dislocate these conventions.

Benjamin notes that the variety and constant flux of urban experience, the complexity of conceptual relations awakened by the urban labyrinth, shocks expectations. This also leads to the undermining of order, the dislocation of conventional dualisms, including the polarity of tradition and modernity. The city is both old and modern, not only a place filled with memory but also the center of social transformation. The city does not merely evoke the past. Urban images may also gain new power and purpose, and may evoke and create history as present and future (Gilloch 1996). In this respect, the city is a space for the play of possibility. The detachment of the modern city from traditional spatial practices and representations frees the vestigial symbolic potency of its auratic objects. Instead of being organized to sanctify traditional social relations of domination, urban images (whether objects or social practices) exist as residues and fragments of social memories, dreams and aspirations which can be applied to the task of social transformation through the creation of new myths (Buck-Morss 1991). In relation to conventional social behaviors, urban public spaces in general are profane rather than sacred. The city streets are promiscuous, permissive, a quality which Benjamin characterizes through the figure of the prostitute (Brown-May 1998). This freedom which the city inspires can be likened to the rule-bending and rule-breaking of play.

It is in the notion of the modern city as the site for the rediscovery, transformation and redeployment of mythic images that Benjamin's analysis of urban experience can be linked with his various observations on play and games. A rich material and symbolic world remains available for (re)discovery and creative use. One form of urban play, initiated by the Surrealists, is wandering, free from goals, compulsions and inhibitions, in a heightened state of distraction. Allowing oneself to be led by fate or caprice, one can lose oneself and one's way in the labyrinth of the city, and can encounter both familiar and unfamiliar objects without necessarily having an instrumental purpose for them. Such activity allows the rediscovery of the world as both old and new through chance exposure to what Benjamin terms 'dialectical images', images perceived in fragments, detached from conventional meanings, which could arouse unfamiliar and contradictory juxtapositions of concrete reality, meanings and memories in the viewer (Buck-Morss 1991). For Benjamin, the practice of wandering is personified by the *flâneur* wandering the streets in a casual but alert manner. The aim of the Situationists in their practice of the *dérive* (or 'drift') was similarly to 'drop their usual motives for action . . . and let themselves be drawn by the attractions of the terrain and the encounters they find there' (Debord 1996c: 22). The *dérive* encouraged situations: the bringing together of aspects of the city which were previously separated in time and space. This convergence created temporary changes in social conditions (Lefebvre 1997a). The Situationists' aim was to understand 'the urban environment as the terrain of a game in which one participates' (Sadler 1998: 120, citing Author unknown 1996b: 84). Their *dérive* highlights that wandering is not purely a matter of chance, but can also be a practice of intentional experimentation, drawing upon the dynamism and the potential for play which was latent in the urban milieu.

In modern urban space, social boundaries, cues and conventions can be recognized but also disregarded and transgressed (Lyman and Scott 1975). Benjamin saw play, like cities, as 'both mythic and demythifying' (Gilloch

1996: 84). The city creates conditions for play because, like play activity itself, it situates objects in new, unconventional relationships, it enhances the recognition of connections which are not about instrumentality or power. It is a center of possibilities which become realized through the decoding and recoding of its images and practices. One way to free mythic images from their status as spectacle and commodity is to intentionally misrecognize their exchange values. The Situationists sought to open up the possibilities of new relationships between social images, to unlock their mythic power, through the process of *détournement*, 'the hijacking of commodities (that carry with them a prescribed reading or utility) into heavily coded, unfamiliar contexts. In a word, *détournement* is the reterritorialisation of the object' (Ball 1987: 34). The concept of *détournement* suggests 'detouring, deflection, and the sudden reversal of a previous articulation or purpose' (Ball 1987: 32). *Détournement* is a 'subversive plagiarism' (Plant 1992: 88). While it evokes familiar meanings, it undermines their authority, both by turning them against themselves, and by denying any certainty of meaning. In doing so, it makes possible the reclamation of lost meanings and 'reveal[s] a totality of possible social and discursive relations which exceeds the spectacle's constraints' (Plant 1992: 87). As such, *détournement* offers a critique of the symbols through which people make sense of everyday life. While '*détournement* characterized the upsetting of relationships with people, cities and ideas [through] games, *dérives* and constructed situations' (Plant 1992: 89), *détournement* can also be understood as an analysis of the way urban space functions in people's everyday perception, reterritorializing every image in new, uncommodified and often irrational relationships.

Contrasting the fetishization of the exchange value of the commodity, Benjamin describes child's play focusing on the waste and byproducts of the adult world which can be found in the urban landscape (Gilloch 1996). Such objects are already freed from their commodity status. In play, these objects are brought together in new intuitive relationships through a process of 'playful (re)construction' (Gilloch 1996: 88). Such relationships may arise from the recognition of similarities among objects or places which are formal rather than instrumental, and this mode of perception becomes possible in a state of distraction, standing outside received myths of origin, purpose and value. Urban space and its symbols are perceived in a state of distraction, outside the focus of people's vision and outside instrumental frames of reference (Savage 1995; Gilloch 1996).

Another linking theme between cities and play is the richness and heightening of sensory experience, the closeness and concreteness of urban experience. Unlike other auratic forms of art, which are experienced through concentration, the city assaults all the senses continuously, awakening a wide range of meanings and desires. The Situationists saw the *dérive* as a means to knowledge about what they called psychogeography: 'the study of the precise laws and specific effects of the geographical environment, consciously organized or not, on the emotions and behavior of individuals ... the sensations they provoke' (Debord 1996a: 18–20). Like Benjamin they believed that such knowledge came to people in a state of distraction while wandering, and that large cities were particularly conducive to this kind of distracted attraction.

The auratic nature of cities also means that they retain a certain sovereign 'otherness' in relation to the viewer and user. Objects of play are likewise approached with a sense of reciprocity, and not with an instrumental or fetishistic attitude. The player avoids predetermined hierarchies of value, and this extends to their own relation to things perceived. The spontaneity of exposures and interactions also denies the possibility of instrumental conquest (Gilloch 1996). A third linking theme is the permissive freedom of the anonymous city streets, which can be likened to the rule-bending and rule-breaking of play.

In summary, the city inspires play because people's movements and perceptions within it constantly arouse a wide range of meanings and memories which do not sit tidily within conventional expectations and trajectories. These loose and unfamiliar phenomena can be employed to create new experiences and new contexts for action.

Benjamin's perspective on how people perceive meaning in urban space, which is very much focused on modern consumer society, provides a strong contrast to the many critiques which highlight the spectacularization of urban space today: the contrivance of architecture, urban planning, public art and civic ceremony to carefully channel the public's desires for excitement, exoticism, freedom and awareness of identity into pre-packaged images and increasingly passive forms of leisure, with the aim of serving instrumental agendas of private profit and social order (Harvey 1989; Sorkin 1992; Gottdiener 1997; Hannigan 1998). Lefebvre (1991b) suggests that this domination of spatial representations by capital and governments, achieved in part through the wholesale reorganization of social space and the design of separate leisure landscapes, is just as important as the domination of other, more 'concrete' means of social production.

But just as with spatial practices, the *oeuvre* of spatial representations is not so easily subsumed to functional ends. Unlike television and other mass media, urban space is a representational medium through which everyone's social life is lived, where its values are continuously being both read and written, often in creative and unexpected ways. Playful acts show people's continued capacity for the invention, discovery, appropriation, contestation, reappropriation and expansion of the meanings that urban spaces can convey. Because social behavior in public is not purely and simply functional, it can be an active, interpretive and expressive response to meaning. The complexity of the city and the diversity of its users mean that there are often contradictions and tensions between meanings received and produced.

The city and everyday life

When Lefebvre and Benjamin describe the concrete and symbolic dimensions of urban experience, and the impact of capitalism upon this experience, they are focusing upon the plane of individual actions and responses. The terrain of their investigation is things that happen to ordinary people and what ordinary people do. Where their discussion of practices and perceptions has focused on defining urban life in contrast to rural or industrialized life, the concept of everyday life defines a particular philosophical perspective on what the structure and content of urban life is. It provides a useful framework within which to understand how various kinds of social activities and values relate to each other within an individual's life and, correspondingly, the ways in which the city organizes these experiences in time and space.

Lefebvre uses everyday life in the same sense as he uses the concepts of use value, urbanism and *oeuvre*, to include the full scope of social acts, in contrast to the ordered rationality of instrumentally productive work. The concept of everyday life addresses several general themes. It attempts to embrace both concrete experiences and conceptual abstractions. The activities of a life are viewed as an undifferentiated, uncategorized totality, and this totality is argued to have a collective style which is locally and historically specific. Everyday life is claimed to have depth, immediacy and authenticity. The concept emphasizes life's cyclical and repetitive nature, its attachment to seasons and to tradition in general. The concept of everyday life allows a range of sociological oppositions and tensions to be brought together in a temporal and behavioral framework (Lefebvre 1991a, 1997b). But this is not to say that the complex of social pursuits which make up everyday urban life is itself tidily resolved. In their great diversity and changeability, 'the social needs inherent to urban society' are both complementary and contradictory:

> Social needs ... opposed and complementary ... include the need for security and opening, the need for certainty and adventure, that of organization of work and of play, the need for the predictable and unpredictable, of similarity and difference, of isolation and encounter, exchange and investments, of independence (even solitude) and communication, of immediate and long-term prospects. The human being has the need ... to see, to hear, to touch, to taste and the need to gather these perceptions in a 'world'. To these ... can be added ... the need for creative activity, for the *oeuvre* ... for information, symbolism, the imaginary and play. Through these specified needs lives and survives a fundamental desire of which play, sexuality, physical activities such as sport, creative activity, art and knowledge are particular expressions and moments, which can more or less overcome the fragmentary division of tasks.
>
> (Lefebvre 1996: 147)

Lefebvre's listing of numerous oppositions suggests that the needs which are served by urban life are not purely instrumental or normative, and that these needs are not pursed in a coordinated and efficient manner. Uncertainty and disorganization are, he argues, just as important to life as their opposites. He also states that 'ambiguity is a category of everyday life', refuting the idea that the full scope of different social needs can be adequately catered for or even categorized (Lefebvre 1991a: 18). The Situationists further note that social laws and human needs and habits are not fixed, but change through history (The Lettrist International 1996). Urban conditions make possible a wider range of lifestyles and they also help to increase the diversity of everyday life as it emerges through social processes such as play.

Lefebvre's and the Situationists' critiques of the modern world center around the total occupation of everyday life by capitalism, 'the concentration-camp organization of life' (Author unknown 1996a: 118). Even non-productive sectors of the economy, such as leisure, aesthetics and urbanism, are reorganized to serve the reproduction of the social relations of mass production and consumption and instrumental exchange (Martins 1982). Lefebvre (1991a) draws upon Marx's analysis of the commodity to explain the implications of capitalist reorganization for everyday life and social practices. The key theoretical concept he employs is alienation. Under the terms of capitalist exchange, places, objects and practices which at one time may have formed an enduring core for the individual's sense of themselves and their place in the world have become external to them, detached. In particular, 'the mysterious, the sacred and the diabolical, magic, ritual, the mystical', which were originally the most immediate and intense aspects of everyday life, have become demoted and displaced under capitalism (Lefebvre 1991a: 117). Benjamin's writings also document the disenchantment of everyday life in the modern metropolis: 'It is here that play is transformed into toil, curiosity into fetishism, reciprocity into tyranny, spontaneity into drudgery' (Gilloch 1996: 91). The Situationists go so far as to claim that 'urbanism makes alienation tangible' (Vaneigem 1996: 127).

Lefebvre's discussions of everyday life and urbanism both center around the contradiction between use value (everyday life and the city) and exchange value (modern industrial production) and the dialectical tension between them (Lefebvre 1971, 1991a). This leads to the oppositional defini-tion of everyday life as what remains of social experience after specialized activities, in particular instrumental labor, have been disregarded (Ball 1987). The hours of a life become fragmented, isolated into pledged time for labor, compulsive time which is filled by the demands of social reproduction, and free time which is used for 'leisure'. Lefebvre's analysis of everyday life in modern society focuses largely on leisure because it is defined as something other than instrumentality and obligation, and hence it reveals something of the broader character, objectives and needs of life. Lefebvre perceives leisure in a dialectical relation with instrumental work. Industrialized production and consumption meet certain human needs, but they also create new needs. People consciously and continuously reshape their leisure activities in relation to the provisions and demands of work. In this way leisure is 'the critique of everyday life from within: the critique which the everyday makes of itself, the critique of the real by the possible and of one aspect of life by another' (Lefebvre 1991a: 9).

There are three aspects to Lefebvre's dialectical characterization of leisure. The first two are that leisure is one distinct part of everyday life, and that it is a critique of life's other dimensions. The third important aspect of leisure is that in its critique it 'reflects' other dimensions of life. The forms leisure takes draw upon the conventional forms of everyday social behavior (Simmel 1950). The first aspect of leisure, its definition as a distinct realm of life, emphasizes the idea of a separation or liberation from other aspects of everyday life. Escape can be spatial, where people physically distance themselves from instrumentality. Lefebvre uses the café as an example of an urban leisure space which provides a liberation from professional and familial structures. He employs the following definition of the domain of leisure:

> An occupation to which the worker can devote himself of his own free will, outside of professional, familial and social needs and obligations, in order to relax, to be entertained or to become more cultivated.
> (Lefebvre 1991a: 32 fn52)

One way to achieve escape is through distraction, which links back to Benjamin's phenomenological observation that the city is both experienced in a state of distraction and also distracts people from teleological actions and prescribed meanings. But the separation between play and other everyday actions may be an issue of content and not just form. Any activity may have the appearance of being everyday, yet not be rational: it may not meet goals, address material concerns or comply with role responsibilities. The fact that leisure rids life of its typical content underpins the observation that play is primarily studied as a matter of form or style (Lefebvre 1991a). It makes more sense to examine the playfulness of any activity, rather than attempting to definitively categorize it as play or not.

Various forms of relaxation, leisure activities characterized by passivity, are ways that people release themselves from the pull of instrumental social purposes, both formally and in terms of content. Benjamin suggests that the passive reception of images such as movies allows people to fantasize about other ways of living, distant from reality. While leisure can be passive, Lefebvre points out that this potentially establishes an alienation from full engagement in the richness of everyday life, that it is particularly susceptible to commodification. He notes that some other kinds of leisure activities are more active and involved, requiring exertion and the development of specialized technical skill or knowledge. This active, intense, engaged side of leisure has many themes which parallel Lefebvre's depiction of urban life generally, such as the importance of encounters with difference and the need for appropriation of space through use. Hence it is highly likely that urban play assumes many active forms.

The different forms of leisure arise in response to different social needs; needs which are not met by the social conditions set by capitalism as well as needs which are a direct product of those conditions. As the Situationist Constant observes, 'once the functions are established, they are followed by play' (Constant 1997: 110). Passivity is a response to instrumental compunction, while involvement is a response to the alienation of commodified processes of exchange. Both modes of leisure attempt to broaden the scope of human experience, to test and stretch the specific constraints set down by mass production and consumption. This is the second aspect of leisure: its role within everyday life as dialectical critique:

> leisure appears as the non-everyday in the everyday. We cannot step beyond the everyday ... There is no escape. And yet we wish to have the illusion of escape as near to hand as possible. An illusion not entirely illusory, but constituting a 'world' both apparent and real ... quite different from the everyday world yet as openended and as closely dovetailed into the everyday as possible ... Thus is established a complex of activities and passivities, of forms of sociability and communication ... they contain within themselves their own spontaneous critique of the everyday. They are that critique in so far as they are other than everyday life, and yet they are in everyday life, they are alienation ... Thus leisure and work and 'private life' make up a dialectical system, a global structure.
>
> (Lefebvre 1991a: 40)

Lefebvre writes at length about various aspects of this critique of everyday life (Lefebvre 1971, 1991a, 1997b). Leisure is an expression of free will, a critique of compunction and the predetermination of one's lot in life. Leisure admits exposure to a richness and closeness of experience: tactility, sensuousness, even eroticism and vulgarity.

Another theme Lefebvre addresses is the critique of economy and efficiency, the rational expenditure of energy. Objects and actions are treated as valuable in themselves, and are not treated as instrumental to other objects and actions. Because leisure's aims are something other than economy, it can also be excessive, using up resources without renewing them. Much leisure is based around intentional loss or wastage. In other cases, leisure centers on activities whose outcomes are indeterminate, subject to fate, and thus leisure also critiques the myth of progress through work, of making sacrifices today for later gains (Lefebvre 1991b). None of this is to suggest that leisure, as a critique of work and other instrumental, alienated social relations, is itself completely free from the effects of fragmentation, homogenization and revalorization by capitalism. Leisure, like work, is regulated, and it is practiced with varying degrees of restraint. It has private and public sides, and it inevitably serves purposes beyond itself. Lefebvre notes that some forms of leisure are difficult to distinguish from other aspects of everyday life. This is because leisure is not itself 'other', not a fully autonomous activity like dreams, art and philosophy (Lefebvre 1991a). It is a critique of everyday life from within and by those who live it. As a critique, play both illustrates and seeks to promote those possibilities of urban life which are endangered by instrumentality. Lefebvre sees industrialization and urbanization as a 'double' dialectical process: each critiques and provides grist for critique by the other. He argues that industrialization is an instrumental enterprise whose ultimate goal and justification can only be 'a fruitful urban life' (Lefebvre 1971: 47).

Edensor suggests that the city can be a heterotopic space, 'an alternative system of spatial (dis)ordering where transitional identities may be sought, sensual and imaginative experimentation indulged' (Edensor 1998: 219). He argues that western tourists visit Indian streets as a form of leisure which expresses 'the contemporary need to reinstate desire, disorder and unpredictability into life' (Edensor 1998: 217). Playful activities which occur in urban public space often arise as a dialectical critique of the stability and rationality of much of contemporary urban life; such activities are not absolutely spontaneous, voluntary or creative. Their diversity is stimulated and given shape by the complex *habitus* of everyday urban social life (Bourdieu 2000). Playful actions are shaped by the dispositions of actors, including their comprehension of the social context, their expectations, the desires that motivate them and their bodily competencies. Individuals draw upon this schema of dispositions as the basis for their adaptation to new and diverse situations (Bourdieu 1977). Like Lefebvre, Bourdieu highlights that *habitus*, the conditions which people encounter in their everyday lives and their creative, playful responses all exist in a dialectical relationship. Individual playful experiences lend meaning, form and potential back to people's objective situation, expanding the possibilities of their everyday life. Play is at one and the same time a product of the process of everyday life, a contradiction of that process, and a producer of it. De Certeau characterizes society as consisting of both foreground practices and institutions which give stability, and 'innumerable other organizing discourses which exist in a state of tension': these discourses 'refract' the foreground order, 'bringing new meanings and practices into play' (de Certeau 1984: 48). Focusing on play foregrounds these practices, the 'structuring' structure of *habitus* under urban conditions.

Some leisure activities offer a critique of everyday urban life through an inversion or heightening of the intensity of more alienated aspects of that life (Lefebvre 1991a). Inversion and intensification in leisure operate as critique when they stand in spatial or temporal opposition to the everyday. Such forms of opposition include dialectical images and places set apart for play. Although the experience of such playful sights and sites is temporary, the most interesting aspect of the temporal dimension of leisure is that it implies the possibility of the inversion or intensification of life within a given space, and not only apart from it. This may lead to the critique of meanings that have become fixed in space: for example, through the profaning of sacred space (Lyman and Scott 1975). The idea of play as something which transforms everyday experience within everyday spaces is one illustration of Lefebvre's contention that urban space is multifunctional. People appropriate spaces for use as they see fit, and uses overlap.

In order to critique contemporary life, its focus on progress and its aliena- tions, one way that leisure emerges in the guise of everyday activity is by drawing upon the memories and traditions of everyday life in the past and ideal visions of future society (Ball 1987). Benjamin's analysis of child's play suggests ways that play engages both the past and future in a dialectical confrontation (Gilloch 1996). On the one hand, play is repetitive. Play activities express the eternal recurrence of society's aspirations in each new generation. It is grounded in imitation of both the social practices and the material world which surround the child. Play is 'an archaic, magical mode of relating to things' and to practices, which precedes the instrumental and fetishistic life of adults (Gilloch 1996: 86). The child collects and operates upon fragments of the past loosened from predetermined social hierarchies and values.

Play also contains utopian impulses. It is non-exploitative and non- hierarchical. Play is subversive of social order and the mythologies which sustain it. The disruptive capacity of play is the opportunities it presents to unravel the mythic *from within*. This dialectical potential is illustrated by Benjamin's description of his own play as a child as aiming 'to renew the old by making it my own' (Benjamin 1974: 286). Through play the 'old- fashioned' is 'rescued', 'reassembled' and 'redeemed' (Gilloch 1996: 88–89). This is true not just of objects, but of traditional practices, whether games or serious social relations. Benjamin draws on the example of Fourier's utopian vision, where children's play provides a model (i.e. a form) for all human relations, and for the expression of passions (Gilloch 1996). In this context, play critiques not human work per se, but its exploitative nature.

Lefebvre, Benjamin and the Situationists all highlight that everyday life includes something more than regularity, obligation and calculation in the pursuit of rational objectives. Their arguments refute the common association of the adjective 'everyday' with 'mundane' and 'habitual'. Social practices, perceptions and needs continuously develop dialectically, through a critique of the concrete and the rational by the possible and the desirable. The diversity and creativity of practices, the demythification of perceptions, and the critique of instrumental need within everyday life all give rise to the transgression of social constraints, the exploration of the social world and of one's own body and imagination.

The density and diversity of the city provide a stimulus and a milieu for this exploration. The continuing specialization of tasks and differentiation of spaces also continuously expand human potential, often in unexpected and improbable ways. The publicness of space and people's anonymity to one another encourage the development of roles and masks and encourage the expression of self. The surplus wealth which is a product of the city's diversity makes possible non-instrumental interactions, and the complexity of urban social space also stimulates such interactions. The disorder of symbolism in the city reawakens memories, demythifies them, and arouses the imagination. All these conditions can potentially override social order and control. The experience of urban space is characterized by multiplicity, ambiguity and contradiction, the unpredictable and the unfamiliar. In these ways, urban public space provides a special realm for play.

Chapter 2

Play and the urban realm

Any attempt to define play runs into the question of whether clear distinctions can be drawn between play and other social practices. Play is typically understood in terms of oppositions. It is contrasted with long-term purposes, productive work, and serious consequences (Goodale and Godbey 1988). Such oppositions have a role in framing people's beliefs and their actions; however, the definitions, intentions and effects of play remain varied and imprecise. Most definitions of play themselves remain 'at play', continuously binding or unraveling. What makes something play and what play 'means' to culture continues to be redefined through changing social practice. The strength of the concept of play relies on the binding together of many different social conditions which people may understand as play, but which cannot be collectively defined by any firm set of rules or boundaries. Because play is not a distinct, discrete set of activities, but rather a characteristic which is present to varying degrees in many different kinds of human behavior, it is necessary to look at play from multiple perspectives, drawing together threads of analysis.

Scholars and players use the term 'play' to describe a great variety of practices and objectives. The meaning and purpose of play differs between individuals and between situations. Play is always a rhetorical construction, and the reasons why someone chooses to use the term 'play' to describe a certain range of behavior depends on their wider values and objectives (Sutton-Smith 1997). Very frequently, play is used to provide a contrast to other aspects of behavior – what is done and how and why it is done – although the opposition itself has varying focus and dimensions. In general terms, 'play' is used as counterpoint to behavior which is 'normal' – everyday, conventional, expected, calculated, practical, constant. Which impacts of play are noteworthy depends on professional interest. Play is in some way unusual, special and different, either in form or in outcome. In this book, play stands principally in contradistinction to people's assumptions about the everyday functionality of the urban built environment. It is a rhetorical device to focus attention on uses of public spaces which are not practical and other than what the spaces were designed for. The definition of play that follows thus does not seek to be exhaustive, but rather to focus on four interrelated ways in which playful behavior can be experienced as an escape from other aspects of everyday life in the contemporary city:

- play involves actions which are non-instrumental;
- there are boundary conditions and rules which separate play from the everyday;
- play involves specific types of activities through which people test and expand limits (competition, chance, simulation and vertigo);
- play in the city very often involves encounters with strangers.

The analytical concept 'play' is most often applied to the experiences of children. Play is seen as largely opposite to the behavior of adults. Children's playful behavior in cities has been examined from a variety of critical perspectives: historical, sociological, ethnographic and autobiographical (Lynch 1977; Nasaw 1985; Dargan and Zeitlin 1990; Benjamin 2006). Such focused studies emphasize the many significant differences between children's experiences in cities and those of the population as a whole. To understand the breadth of potential offered by the urban condition, it is necessary to examine how the population as a whole uses public space for play. However, there are a number of substantive reasons why children's play, and existing research into children's play, is generally disregarded in this book.

Children's play occupies a more narrow range of behavior than the play of adults. Play is just one component of the complex social existence of working adults, and one that is rarely analyzed. Adult play is not merely a remnant of childhood forms; indeed, 'the full variety of play forms only appears with the achievement of a certain maturity' (Mouledoux 1977: 52–53). Adults may play less often than children, yet adults have knowledge, abilities and a freedom of action which permit them to play in times and places and in ways which are not available to children, and this is particularly true of public spaces in the inner city. It is also only within the complex context of adult social life that play's dialectical qualities become apparent. The density and diversity of urban settings intensifies the tensions and contradictions between the serious world of adults and their playful escapades. Thus adult play provides far better illustration of the transformation of everyday life and of lived space into new experiences and new forms. It is the play of adults which can lead to a reconsideration of the ways in which urban space might stimulate and facilitate unexpected and impractical behavior, and how space can be utilized for escapes from serious meanings and uses and to critique the normal social order.

Another limitation of examining children's behavior is that theorizations of children's play tend to circumscribe the freedom, creativity and diversity of human agency rather than open it up. Because children's skills and ambitions are limited, their play is of only certain kinds. Following the work of Piaget, the field of developmental psychology views children's play as a process which aids learning and socialization (Spariosu 1989). For children, play lies at the center of their experience of the world. Play is the primary 'function' which they are supposed to pursue. Children's play activities are generally accepted, even encouraged. They are also supervised. In these ways, child's play generally reproduces the *habitus* which defines childhood itself. A concomitant belief that play is a sign of immaturity leaves Piaget's theory unable to explain how and why adults play (Huizinga 1970). While play can bring about desirable social outcomes, '[it] does not take place because it is functional or useful' (Goodale and Godbey 1988: 174). Interpreting play acts in terms of their utility ignores those intriguing dimensions of play which are most characteristic of urban life: the stimulus of accident and caprice, deliberate exposure to difference and to risk, and the potential which such experiences provide for the continual diversification of social practice.

Benjamin has a distinct theorization of the relationship between children's play and urban experience. He suggests that 'the child in the city is a figure of utopian dreaming' (Gilloch 1996: 91). Benjamin uses child's play as an analogy for various social desires and ideals which are frustrated by the instrumentalism and spectacularization of urban life. Benjamin's view of the potential for adult play in the city is generally pessimistic: 'To recognize yet disregard the invisible boundaries of the cityscape – this is the desire of the child and the regret of the adult' (Gilloch 1996: 85). Here Benjamin's thinking tends, somewhat ironically, towards essentialism and ideology: he reifies children's play as 'archaic' and 'magical'. Benjamin's characterizations of the city and play also have a very contrary dimension, examining how they subvert prohibitions; their capacity for innovation; the new juxtapositions and tensions they constantly produce. As Benjamin noted, 'playfulness and dreaming are part enchantment, part disenchantment, of the adult world' (Gilloch 1996: 92). The play of adults in urban space can enable a re-enchantment of their world; taking advantage of conditions under which toil may be transformed into play, fetishism into curiosity, tyranny into reciprocity, and drudgery into spontaneity (Gilloch 1996).

A second major distinction which is important is that between leisure and play. Leisure and play are closely related concepts, and critical analyses of leisure can certainly aid a better understanding of play. Leisure can most broadly be understood as the luxury of passing time free from compulsion, and in particular from the need to engage in productive activities (Goodale and Godbey 1988). Nevertheless, leisure is a rather precise social construct which is codified in particular practices, and which tends to be demarcated within special spaces and times. It carries connotations of rest and recuperation; of bodily passivity, escape from the busyness and tensions of the social world, and attention to the private life of family and self (Rojek 1995). These circumstances renounce the diversity, intensity and complexity of the city, rather than embracing them. Play, by contrast, is a concept which highlights the potentials of urban experience for promoting and framing active, creative, and above all public behavior. While play can often arise in a context of leisure time, it does not depend upon it. Indeed, the social segregation and ordering of leisure serves to undermine the playful potential of every social experience, by limiting the prospects for confrontations and creative engagements between necessity and caprice, intention and accident, productive effort and waste.

Play is defined not merely oppositionally, but dialectically within everyday life, and in a dialectical engagement with the inherently contradictory 'social needs' which life experiences reflect. Practices of play are a critical response to specific historical sociospatial circumstances; this response can most easily be encapsulated by the idea of escape. Lefebvre (1991a) identifies the dialectical tension underlying escape through play: escape is impossible, illusory, but this illusion in itself constitutes a perceptual and social reality. The concept of play embraces a variety of ways in which people test and transgress the limits of their social existence. In terms of play within the urban public realm, Lefebvre's (1991b) critique of modern city planning suggests that play means encounters with difference, encounters which contest the fragmentation and alienation of contemporary social experience.

Non-instrumentality

Play is presented in western metaphysics in opposition to seriousness, morality, and productive work, and the social power relations these value structures help reproduce (Spariosu 1989). Play is quite contrary to these values: 'Play is spontaneous and creative, a counterpoint to the tedium and exploitation inherent in instrumental labor. It is the domain of freedom from compulsion' (Gilloch 1996: 84). Play activities are irrational because they are not shaped around conscious, preformulated ethical and pragmatic goals. Play often runs against orthodoxy, ignoring the systematic organization of human activity, and transgressing the boundaries of seriousness, including taboos. Play illustrates Lefebvre's (1971) view that the practices of everyday life are far richer and broader in scope than rationalism and morality can explain and provide for. Lefebvre's (1996) use of the term *oeuvre* conveys the sense of everyday life as being a work in itself, not a series of means toward predetermined ends. Lofland (1998: 121) writes about 'the sacrilegious frivolity of uncontrolled play' in urban space: 'In the public realm . . . the unquestioned virtues of sobriety, industry, rationality, diligence and so forth are not only challenged, they are discarded.'

Play's opposition to instrumentality embraces the issues of purpose, functionalism and productivity. Play is purposeless, free from the 'pitfalls of teleology' (Spariosu 1989: 90). People accept the instrumental organization of much of everyday life as being necessary to meet human needs. Gratification is deferred in the name of future pleasure or comfort. When viewed in terms of these kinds of conventional social purposes and needs, play often seems irrational. Play actions thus offer a critique of conventional understandings of purpose and need, calling for a different way of thinking about these matters (Rojek 1995).

Instrumental social practices and relations themselves present contradictions, because while they meet needs, they also engender alienation from aspects of the self and from other people. Hence they establish new needs dialectically. Although human needs have a foundation in biology, they are themselves a product of social life and human consciousness (Lefebvre 1991a).

Rationality suggests people should pursue optimal fulfillment of needs within a given ethical framework. Yet it is precisely when people's actions are not locked into the service of future goals that their actions are free to explore human values, and can thereby constitute meaning: 'What [life] allows in the way of order and reserve has meaning only from the moment when the ordered and reserved forces liberate and lose themselves for ends that cannot be subordinated to anything one can account for' (Bataille 1985: 128). Thus play, as a pleasurable end in itself, arises as 'liberty from every social, erotic and psychological constraint' (Castle 1986: 53). Ethics sets rules and boundaries which conserve and protect social structure. Play ignores these boundaries. Nietzsche argues that human behavior is not inherently based in ethics, and in fact can be 'justified only in aesthetic terms' (Spariosu 1989: 80–81). As Nietzsche suggests, the whole world is 'eternally selfcreating, the eternally selfdestroying . . . "beyond good and evil," without goal . . . without will' (Spariosu 1989: 89, citing Nietzsche 1968: 549–50). What Lefebvre proposes is the importance of pursuing a richer understanding of the self and one's needs and desires, outside the framework of rationality and ethics: 'The more needs a human being has, the more he exists. The more powers and aptitudes he is able to exercise, the more he is free' (Lefebvre 1991a: 161). This sense of play as the exploratory pursuit of pleasure is captured in the figure of Baudelaire's urban *flâneur*:

> The *flâneur* is defined as a constant seeker of impressions and stimuli . . . But he does so in a spirit of idle curiosity, without any object of learning anything or reaching understanding . . . the *flâneur*, then, cultivates polymorphousness and discontinuity in leisure . . . He makes a virtue out of idleness and values the senses above reason.
>
> (Rojek 1995: 91)

Play is defined as freedom from the instrumental pursuit of social purposes. But this freedom does not arise naturally through environmental and technological opportunity. Freedom is not wholly defined by an absence of power, by finding gaps within the instrumental structure of everyday life. Huizinga (1970) notes that play is freedom. Practices of play constitute, rather than merely reflect, freedom. Freedom takes place in dialectical relation to power. The more a form of play is defined as immoral, outlawed and restricted, the greater its attraction as an escape from, and confrontation of, the social order (Rojek 1995). The boundaries of freedom get established through action and reaction, rules and the transgression of rules. Everyday actions in the city are so many and so various that they resist any totalizing control or any overall schema.

In terms of function, play runs contrary to the idea that each human action is designed and performed to effect a predetermined change in the material or social world. Play behavior is gratuitous, it does not achieve ethical or material social outcomes. Benjamin's interpretation is that acts of play are timeless, repetitive rather than developmental, removed from the myth of history as progress (Gilloch 1996). Bauman (1993: 171) concurs that 'to play is to rehearse eternity . . . Nothing accrues, nothing "builds up", each new play is an absolute beginning'.

Writing about the social value of the city, Lefebvre (1996) notes that urban play is part of a broadened conception of human needs. He also points out that 'functions', by which he means the full diversity of human actions, do not just meet specific needs, but are themselves needs, and also that functions of play help to develop the concept of human needs. To contrast with capitalist relations of instrumental exchange, which fragment life into specialized techniques, locations and spheres of meaning, Lefebvre (1997b) presents his conception of the *oeuvre*, the totality of undifferentiated everyday urban life in a particular sociohistorical setting. Social practices are diverse, contradictory and ambiguous. Practices of play exemplify the transfunctionality of social life.

A third way of understanding how play resists instrumentality is through its critique of the idea of productive output. In contrast to the rational organization of human activity based on production, everyday life presents 'the opposite thesis . . . according to which waste, play, struggle, art, festival – in short, Eros – are themselves a necessity' (Lefebvre 1991b: 177). Play, as part of the 'Dionysian side of existence – excess, intoxication, risks (even mortal risks) – has its own peculiar freedom and value' (Lefebvre 1991b: 178). Rather than producing material wealth, play consumes; 'Play is an occasion of pure waste: waste of time, energy, ingenuity, skill, and often of money' (Caillois 1961: 5). Play takes advantage of the surplus which exists in the natural world and which is made possible by human work. Excess arises from production which is greater than necessary consumption for reproduction of the organism (Bataille 1988). The accumulation and discharge of surplus energy is a necessary and defining characteristic of living bodies (Lefebvre 1991b). This analysis fleshes out Lefebvre's concept of the *oeuvre* as necessarily combining productive and unproductive acts. In the case of non-productive acts, '[surplus] allows the organism a measure of leeway for taking initiatives (these being *neither determined nor arbitrary*)'

(Lefebvre 1991b: 176, emphasis added). That is to say, the use of surplus is an issue of gratuitous choice.

Bataille (1985) argues that a society is defined primarily by how it chooses to use or 'waste' its surplus, and not by its mode of production. Waste is a path of escape from the cycle of acquisition, productivity and conservation, which leads to an act being understood as an end in itself. The meaning of a culture's various unproductive acts arises in this unrecouped consumption: these acts become the chosen ends for which people willingly subsume practical actions as means: 'luxury, mourning, war, cults, games, spectacles, arts, perverse sexual activity, the construction of sumptuary monuments – all these represent activities which, at least in primitive circumstances, have no end beyond themselves' (Bataille 1985: 118). The intense human significance of such activities highlights that play is 'not time wasted but time filled with profound and rich experience' (Clark and Holquist 1984: 303). Material abundance makes freedom from productive work possible, because instrumental needs and demands are not constraining people's actions. The city is an engine of wealth, and the reserves of surplus energy that the city produces are reflected in the scope or *oeuvre* of urban life. Urban public open space itself is one of the luxuries afforded by excess productivity, but little seems to be known about how people take advantage of this luxury, or how the urban environment in general frames experiences of excess, intensity and exposure to risks. One clear distinction that has been identified is that people tend to do instrumental tasks in public space as quickly as possible, whereas they linger over 'optional activities', if there are public spaces which are comfortable for doing them in (Gehl 1987; Gehl and City of Melbourne 1994). In a sense, good public spaces are always used inefficiently; the space is always wasted.

Productive and unproductive activities have an important relation to each other, making up a 'transfunctional' totality of social practices:

> The human being has the need to accumulate energies and to spend them, even waste them in play ... play, sexuality, physical activities such as sport, creative activity, art and knowledge are particular expressions and moments, which can more or less overcome the fragmentary division of tasks.
>
> (Lefebvre 1996: 147)

Various modes of play escape the production of surplus and the routinization of function in different ways. Leisure, as an escape from work, 'embraces opposing possibilities and orientations', and it has distinct passive and active forms (Lefebvre 1991a: 32). Passive forms of leisure halt the process of production and reproduction. They allow for the dissipation of accumulated resources, in particular the wasting away of time. Passive activities include reading a novel and attending the cinema. Lefebvre uses the same concepts to characterize passive leisure as Benjamin uses to characterize people's general experience of cities: distraction, alienation through the creation of a vacuum, and the illusion of escape from the everyday. Active forms of play, such as hobbies and skills, redirect people's excess productive potential to tasks which are not purely rational. These forms of play involve the exertion of energy or the intellect and the control of muscles. They provide for the rapid and dramatic consumption of resources.

A second way of differentiating between the various forms of play comes from consideration of play as an escape from 'the localization and "punctualization" of activities' (Lefebvre 1991b: 227). Capitalism tends to restructure all social relations, both productive and consumptive, in time and space so as to maximize instrumental functionality. Play can be seen as 'a voluntary departure from the mundane world of involuntary routinization' (Lyman and Scott 1975: 147). This escape also leads in two different directions. Caillois (1961) suggests that all play activities can be evaluated along a continuum between *paidia* and *ludus*.

Play as *paidia* is characterized by diversion, destruction, spontaneity, caprice, turbulence and exuberance. *Paidia* is human will acting without ethical deliberation. This enhances one's awareness of being a cause, a free and active force which shapes reality. *Paidia* is both a refusal to accept limits and a willful transgression of them. It has no civilizing 'function' (Spariosu 1989, 1997). *Paidia* is improvisatory action, an escape from routine which explores other possibilities of social experience and which develops new social forms. *Paidia* is typified by the play of children, who are unselfconscious about their feelings and actions. Yet adults retain an attraction to undisciplined behavior and exposure to risk (Mouledoux 1977).

Ludus is play institutionalized as a game. It follows rules and routines which are purposely contrived to be tedious and arbitrary. Such play is 'a secondary and gratuitous activity, undertaken and pursued for pleasure . . . in a word any occupation that is primarily a compensation for the injury to personality caused by bondage to work of an automatic and picayune character' (Caillois 1961: 32). Subordination of individual will to the rules of *ludus* is imperative. It requires effort, patience and skill. The pleasure of *ludus* lies in the development and mastery of technique, the psychological satisfaction which comes from discovering solutions within a set framework which is external to the demands of instrumental function.

Caillois' concepts of *paidia* and *ludus* highlight that escape from instrumentality and compunction can be found either in resistance to rules or in observance of different, and in many cases more constricting rules. Scientific analysis has generally focused on the instrumental utility of such forms of play as a biological function. *Ludus* has been viewed as an aid to cognition and learning, through the imitation and repetition of physical skills. This spirit of discovery and testing is seen as a higher form of work: its creative role assists humankind's evolutionary adaptation to change (Spariosu 1989). Through play one also acquires social skills such as cooperation and collective discipline. People learn that they are the cause of certain outcomes, but they learn this in a context of control. Peaceful play as *ludus* emphasizes order and continuity. It situates play as a part of the routine rationality of everyday life (Rojek 1995).

Caillois argues that social practices of play follow a general progression from *paidia* toward *ludus*, but he retains a dialectical view of practices of *ludus* as freely determined and non-instrumental. He suggests experiences gained through *paidia* stimulate the desire to 'invent and abide by rules' which 'discipline and enrich' it (Caillois 1961: 28–29). He gives the example of the theatre, where imitation 'becomes an art rich in a thousand diverse routines [and] refined techniques' (Caillois 1961: 31). Caillois suggests that the instituting of rules does not necessarily restrict behavior: 'what to begin with seems to be a situation susceptible to infinite repetition turns out to be capable of producing ever new combinations' (Caillois 1961: 30). *Ludus* allows people to purposefully utilize and develop their skills and knowledge in tasks which are of their own choosing and under their own control.

Drawing on Caillois' work, Mouledoux (1977: 54) concludes that there is both repetition and variation in games: 'every play activity is simultaneously individual and social', as it draws upon cultural knowledge and is adapted to changing context. Indeed, rules of play themselves provide a constant stimulus to diversion, resistance, and transformation through *paidia*. In Wittgenstein's words, 'we play and make up the rules as we go along . . . and . . . alter them as we go along' (Wittgenstein 1958: §83). The notion that there is a continuum between *paidia* and *ludus*, that unstructured play becomes institutionalized over time, and that play forms are derived from serious elements of life, also highlights Lefebvre's (1996) characterization of a continuity of work and play within the *oeuvre* of everyday life.

Boundary conditions

Play activities are distinguished from instrumental labor by a range of physical, psychological and social conditions (Huizinga 1970). Prime among these is freedom: participation in play is by necessity voluntary. Play is elevated to be something more than instinct, rationality or social obligation because 'for the adult and responsible human being play is a function which he could equally well leave alone. Play is superfluous . . . it is never a task' (Huizinga 1970: 26).

A second characteristic is that play is always undertaken with a certain air of disinterest. This is possible because 'it stands outside the immediate satisfaction of wants and appetites' (Huizinga 1970: 27). Play allows people to 'step outside themselves' and their everyday instrumental goals (Lennard and Lennard 1984: 67). Lennard and Lennard (1984) argue that because of the air of disinterest, social relations in play are rarely exploitative. All participants must have freedom of participation; 'In principle nobody can find satisfaction [in genuinely sociable interactions] if it has to be at the cost of diametrically opposed feelings which the other may have' (Simmel 1950: 48).

The idea that play requires a disinterest in everyday life leads to Huizinga's third major distinction between play and seriousness: it is 'a world apart' separated in time and space. This separation allows people to forget their everyday roles, conventions, demands, and restrictions. When considering the case of urban public space, it is perhaps more accurate to say that separations are created for instrumentality rather than for play. Lefebvre's analysis is that the *oeuvre* of everyday urban life is creative and playful, and that rational labor is a special, fragmented domain entered into. Work carves out a rational space and time for itself, with its own special, practical rules and role relations. Through history, elements of natural space and time, their daily and seasonal rhythms, have been regularized, commodified, organized. This can be said not only of work, but also ultimately of leisure: the consumption of all time has become bounded. All boundaries and rules are historically situated, and hence the social structuring of play is continually being redefined by both technological opportunity and social convention.

Many forms of play necessarily occur in places which are physically or socially defined as 'forbidden, isolated, hedged round, hallowed, within which special rules obtain . . . dedicated to the performance of an act apart' (Huizinga 1970: 28–29). The physical boundaries of spaces can be very definite, obvious and determinant of people's actions. This is not always a necessary or desirable context for public play. Physical boundaries are also not always enough to prevent play. Boundaries of various kinds frame limits to human experience; but these limits are part of what get tested through play. Understanding the ways in which urban space both collects people together and separates them, how it shapes the arrangement of serious and frivolous activities, and how it helps supports particular roles for players is crucial for understanding what makes urban space playful.

The social acceptance of the separation of play from seriousness also means the acceptance of rules which structure it. To play successfully requires participants to respect the illusion and the behavioral constraints which set the play world apart from everyday life. Cheating is bad because it involves breaking play's rules of fairness in order to win, but being a spoilsport is worse because it actually destroys the fragile basis of play itself. The need to respect the play world gives rise to the special mood in which people undertake play: simultaneously aware that things are 'not real' and yet willing to believe and participate. Successful play with others relies on 'metacommunication': the message 'this is play' is continuously being communicated between participants, although what is happening may appear serious. The spatial and temporal framing of the message 'this is play' becomes part of this message (Bateson 1987). Thus a place and occasion of play is constituted not merely as a boundary which circumscribes actions, but also as a context of signals which make those actions meaningful.

People choose to participate in play for hedonistic reasons. They receive psychological benefit from the activity as an end in itself, rather than obtaining material benefit as a result of sacrifice and the deferral of gratification. Activities which are not emotionally pleasurable to all participants are not playful. This is not to say that play events cannot inspire emotions such as fear, sadness, tension or boredom. In fact, play is significant in arousing and providing outlet for such feelings when they are otherwise lacking or diminished in everyday life. People enjoy exposing themselves to intense experience, so long as they can choose the limits of the intensity and consequentiality of their exposure. The origin of the word 'play' is in the Old English *plegen*, the meaning of which includes taking risks and exposing oneself to danger of injury or failure. Play includes the freedom to attempt something and to fail (Goodale and Godbey 1988). Taking risks adds strength and depth of people's experience in the world. They know play is not reality, yet within the delimited context of play events they allow themselves to believe and to act as if some aspects of risk are real and large, and to experience the tension and thrill of handling such risks.

The idea that play occurs in a time and place apart suggests tensions. Because play is often stimulated as a response to serious reality, it also often arises in a physical or temporal proximity to seriousness. Such proximity always involves uncertainty; limits are always to some degree unknown. Particularly in urban public settings, people at play are always to some degree in real jeopardy, because they are bodily present, and events are linked in complex ways (Goffman 1982). Not everyone present necessarily has the same understanding of the rules that apply, and the level of different people's involvement in play cannot easily be predetermined or controlled.

A typology of play forms

The various aims and contexts for play underscore that it is a strongly personal, subjective experience. However, rather than psychology, the focus here is on the role of built form in framing a class of spatial events which come under the rubric of playful behavior. Caillois (1961) defines four basic forms of playful activity: competition, chance, simulation and vertigo (his terms are *âgon*, *alea*, *mimicry*, and *ilinx*). This typology gives useful insight into what characteristics make practices of play different to the instrumentality of work and consumption. These differences include the sets of rules and roles assumed by various participants. Each of the forms of activity also illustrates something about what makes play enjoyable, and hence why people might choose to play. Urban settings frame particular kinds of opportunities for each form of play. Urban public space brings together and multiplies the diversity of social life and social values, and thus public play in cities usually combines several forms.

Competition

In competitive play people seek ways of utilizing their knowledge and skills. Competitive play in urban space includes all manner of individual displays and tests of strength, agility, refinement, intellect and allure. Open conflict aims for mastery over others, but the true goal of competitive play is to foster mastery of the self, the testing of human limits. Such playful engagements with difference force people to reach outside the self and overcome the feeling of omnipotence, by helping people understand their limits. Such actions have some playful element: they are exploratory, and not based on existing power relations (Sennett 1971). The playful form of competition differs from combat for survival or success in that 'equality of chances is artificially created' (Caillois 1961: 14). In contrast to 'open' competition, competitive play follows rules which typically restrict the techniques or attributes which players can employ. The presence of a public audience helps to verify and enforce the rules of fairness.

Public play gives each individual the opportunity to demonstrate, to realize and to expand their capacities for excellence. The city brings together people in all their diversity, each with different abilities and trajectories. Competitive displays occur in urban settings which draw potential players together in front of an appreciative audience. Promenading, dancing and simply being seen in public can be competitive when people show off, aware that others are judging them. Physical space and the arrangement of activities are part of the social context which regulates and inspires playful competition. The city frames the widest range of people to compete against and a great diversity of settings for their struggles. Meanings embedded in public spaces can also help shape the contests which occur within them.

Chance

In contrast to competition, the appeal of chance is that it negates the benefit of any kind of effort, experience, or skill. Chance allows people to escape from human rationality by abandoning themselves to incalculable forces (Lyman and Scott 1975). Benjamin's analysis of gambling foregrounds the dialectical relation between chance and instrumentality:

> In refusing to equate labor, time and money, the gambler appears to resist the discipline of capitalist production ... Indeed, the gambler denies the financial importance of gambling itself, purporting instead to relish the game for its own sake ... 'The more that life becomes administratively regulated, the more people must learn waiting. Games

of chance have the great charm of liberating people from waiting' . . .
Gambling is fundamentally immoral . . . To have abundance without
toil is the utopian promise extended by gambling. It denies scarcity and
the need for rational calculation . . . Each game is independent of the
one that preceded it. The world is repeatedly encountered 'for the first
time'. The role of memory is negated.

(Gilloch 1996: 157–60,
quoting Benjamin 1974, vol. V: 178)

Children and animals do not play with chance:

[Being] very much involved in the immediate and enslaved by their
impulses, [they] cannot conceive of an abstract and inanimate power,
to whose verdict they would passively submit in advance of the game.
[In addition] the child is immune to the main attraction of games of
chance, deprived as he is of economic independence, since he has no
money of his own. Games of chance have no power to thrill him.

(Caillois 1961: 18–19)

Chance can be playful when people understand and accept the limits of risk.
Children lack the critical faculties for foresight and objective calculation
which get tested in games of chance. Chance offers a momentary possibility
of breaking free from the predetermined cycle of production and consumption. Chance is captivating for adults because of the tension brought on by
the risk of loss and the possibility of gain.

Urban spatial form creates conditions of chance which engender various
kinds of playful activity. Life in the city is itself a large game of chance. As
Benjamin noted, the public spaces of the city are a labyrinth within which
wandering flânerie exposes people to seemingly haphazard patterns of new
events. Chance encounters in the city provide opportunities for escape from
predetermined and ritualized courses of action. Many social activities in
public are playful because they are spontaneous, derived from dynamic
conditions of the place, occasion and individuals present (Lennard and
Lennard 1984).

Urban life suggests meetings, the confrontation of differences . . . as a
place of encounters . . . the urban become what it always was: place of
desire, permanent disequilibrium, seat of the dissolution of normalities
and constraints, the moment of play and of the unpredictable.

(Lefebvre 1996: 75, 129)

The continuous system of public spaces which constitute a city, people's
freedom of movement through it, and the density and diversity of activity
patterns found there all bring together different people in urban spaces,

and thus increase the likelihood of unplanned and unpredictable social encounters. Chance contacts expand the individual's experience of life, drawing upon the potential of differentiated individual capacities, and this stimulates further diversification, the development of new values and new social behaviors. Thus chance adds to the citizen's freedom from fixed ways of inhabiting the city (Sennett 1971).

Competition and chance are often brought together in play. Spontaneous human interactions in public places are tests of human behavior between individuals who do not know each other's standards or limits. These interactions hence become games about the negotiation of social rules, for example regarding who is the focus of attention. On the other hand, the controlled nature of playful competition suggests the possibility that other uncontrolled external factors may actually determine the outcome of such games. Competition and chance are also the two forms of play which engage with material spatial circumstances and specific social conditions. They represent people's attempts either to take more control over their world or to abandon themselves to the circumstantiality of that world. They best illustrate play's critique of social practices and the ways in which play contributes to the breadth of urban social life.

Simulation

Simulation is that form of play which institutes a dialectic by counterposing perceptions of the imagined and the real. Simulation is the fabrication of a different character or situation. This involves forgetting, disguising or otherwise escaping one's usual self and one's place in the world: 'the mask disguises the conventional self and liberates the true personality' (Caillois 1961: 21). In contrast to instrumental domination through differentiated roles, this mode of play pursues a sense of reciprocity with things in the world, by becoming like them (Gilloch 1996). Simulation includes any game where people pretend; 'its realm is by no means limited to what one person can imitate in another' (Benjamin 1974, vol. II: 210, quoted in Buck-Morss 1991: 266). It may also involve the creation of new meanings or a new reality, and for this reason the term 'simulation' is more appropriate than Caillois' 'mimicry'. Illusion, theatre and spectacle are all forms of simulation, and urban space assists in the practice of each. Urban conditions help to serve people's needs 'to see, to hear, to touch, to taste and the need to gather these perceptions in a "world" . . . the need for creative activity . . . for information, symbolism, the imaginary and play' (Lefebvre 1996: 147). Simulative acts create the impression of the logic of a world, but a world which is never real, because the simulated actions do not carry real consequences.

Of all the modes of play, simulation is most structured around the idea of an audience, although it also includes solitary pretending and imagining. The design of urban space can reinforce the notion that a simulative play event is occurring by structuring roles of participant and audience (Lennard and Lennard 1984, 1995). Many successful urban spaces are enclosed like a room, limiting the view out, and focusing attention on people and events within. Seating and relatively passive activities such as dining are often arranged at the periphery of public spaces, and this encourages people to linger and watch activities in the space. People will use steps, planters and window ledges for the same purpose (Whyte 1988). Doorways and paths which open onto or next to a public space create opportunities for players to enter and leave the scene. The organization of people's activities in and around public spaces contributes to their theatrical quality, by furnishing a broad range of potential actors and framing their interactions. Numbers and diversity of people using urban public spaces maximize the likelihood that there will be something interesting to watch. Where space is open to public use and there is free movement to and from the spaces surrounding it, it is also easy to make transitions between the roles of observer and participant.

The design of public spaces also supports and encourages memory and fantasy: the complexity of architecture and space in a city provides a myriad of settings and props which can catalyze imaginative play. Throughout history, urban space has been used for cultural ceremonies which evoke other historical and spatial contexts that give legitimacy to social practices. Spectacular effects of scale and intensity amplify such messages; in doing so they play with reality. Ceremonies add social meanings to the spaces where they take place, particularly when they are repeated regularly. Through changing practices, the meanings and functions of a place remain open to reinterpretation and innovative use.

Simulation differs greatly from both competition and chance in terms of the control individuals have over the rules which circumscribe play actions, and the contexts of meaning within which they are framed. Public expression through simulation enhances individual freedom, and people's experience of otherness and of the social whole:

> Public places provide the setting and the opportunity to relate in a variety of modes with friend and stranger alike, both as spectators and per-formers. All who participate learn that others like themselves are capable of a rich repertoire of social behavior and reactions, and are gifted with some ability to be joyous and to give pleasure to others. Consequently it becomes less likely to confuse persons with a particular role function. This learning carries over into other settings and more structured relationships.
>
> (Lennard and Lennard 1984: 18)

Unlike competition and chance, simulation is not bounded by precise rules. Simulation requires measures of both reproduction and deception, the invention of new codes of meaning. It gives each individual opportunities to invent roles, which means transcending predetermined social relations and conventions (Sennett 1971, 1974). Simulation relies on the constant substitution of something for something else. In simulation people test the connections between perceptions and the meanings which society has assigned to them. People break these connections willingly, and suspend their disbelief: 'The rule of the game is unique: it consists in the actor's fascinating the spectator, while avoiding an error that might lead the spectator to break the spell' (Caillois 1961: 22–23).

Vertigo

Vertigo includes a wide variety of behaviors through which people escape normal bodily experience and self-control. Caillois characterizes vertigo as 'surrendering to a . . . shock . . . which destroys reality' (Caillois 1961: 23). In vertigo, people 'lose themselves' and are transported to new forms of experience. Some experiences of vertigo come through bodily actions which generate intoxicating physical sensations of instability and distorted perception: 'mad, tremendous and convulsive movements' such as falling, sliding, jumping, climbing, dancing, spinning and moving quickly. These acts 'inflict a kind of voluptuous panic upon an otherwise lucid mind' (Caillois 1961: 25). This category also includes such activities as skiing and tightrope walking: bodily encounters with space which provoke vertigo in the commonly accepted sense. The common feature is 'the voluptuous experience of fear, thrills and shock that causes a momentary loss of self-control', which allows one to step outside normal, stable perception of the world and bodily practice within it (Caillois 1961: 169). The classification of physiological vertigo embraces a wide range of direct confrontations with the physical environment, including any activity which is a bodily competition framed by the risks the environment presents, including height but also scale, speed, and traction. The feeling of vertigo is often achieved with the aid of equipment, which in public space includes skateboards, bicycles and in-line skates.

Because play has to remain separated from seriousness, experiences of vertigo remain pleasurable when they are calculated risks of limited duration. The surrender to shock cannot be total, because 'the danger lies in not being able to end the disorder that has been accepted' (Caillois 1961: 78). Vertigo often involves a tension between the desires for risk and control, creating

> a world without rules in which the player constantly improvises . . . The person lets himself drift and become intoxicated through feeling directed, dominated, and possessed by strange powers. To attain them, he need only abandon himself, since the exercise of no special aptitude is required.
> (Caillois 1961: 75, 78)

However there is also a desire for *ludus* underpinning dangerous vertiginous play like skiing and tightrope walking, when people concentrate their attention on 'training in self-control, an arduous effort to preserve calm and equilibrium . . . to neutralize the dangerous effects of [vertigo]' (Caillois 1961: 31). Rather than uncontrolled motion, such play relies on 'elaborate maneuvers', 'natural creativity' and strength of nerve, the ongoing mastery of one's exposure to dangerous circumstances. For example, an acrobat succeeds 'only if he is sure enough of himself to rely upon vertigo instead of trying to resist it. Vertigo is an integral part of nature, and one controls it only in obeying it' (Caillois 1961: 138). The thrill of vertigo comes about when people control their own encounters with the uncontrolled, the irrational, the extreme and the violent.

A second distinct category of vertiginous play is psychological or 'moral' vertigo. This takes such forms as breaking objects, making loud noises, fighting and pressing oneself into a dense crowd. 'This vertigo is readily linked to the desire for disorder and destruction, a drive which is normally repressed' (Caillois 1961: 24). Although they generate intense sensations, these crude acts focus on the disruption not of perception, but social propriety. They are expressions of individual agency 'in rebellion against every type of code, rule and organization' (Caillois 1961: 157). Through such acts, people seek to experience a sense of release, freedom 'from the burden of memory and from the terrors of social responsibilities and pressures' (Caillois 1961: 51). Practices of moral vertigo frame 'the abdication of conscience' (Caillois 1961: 44). People engaged in playful forms of vertigo experiment with deviance; they 'toy with violence and tease repressed passions' (Darnton 1984: 101). Vertigo most clearly characterizes the Nietzschean idea of play as being outside ethics (Spariosu 1989).

The primal, Dionysian impulse to vertigo remains tempting, despite its real, life-threatening dangers, and despite the regular sublimation of desires into the more organized, 'cultural' forms of play, namely competition and chance. Chance is 'a mockery of work, of patient and persevering labor, of saving . . . in sum, a mockery of all the virtues needed in a world dedicated to the accumulation of wealth' (Caillois 1961: 157). In a similar sense, the exuberance and excessiveness of vertigo persists as a dialectical response to the overbearing physical and psychological security provided by the social order, which alienates the individual from direct engagement with the material world and from a sense of agency within the social world. This is borne out in Benjamin's view of children's play as disruptive and subversive, overcoming prohibitions and inhibitions (Gilloch 1996), and in Caillois' evaluation of why Huizinga ignores games of vertigo:

> He no doubt holds them in disdain, because it seems impossible to attribute a cultural or educative value to games of vertigo . . . They are

regarded as destructive to the mores. According to a popular view, culture ought to defend itself against seduction by them, rather than profit from their controversial revenues.

(Caillois 1961: 169–70)

The controversy regarding the revenues of play as vertigo is that it is linked to conceptions of value and need which are non-instrumental: 'Aberrant disciplines, heroic feats accomplished to no purpose or profit, disinterested, mortally dangerous and useless, they are of merit in furnishing admirable witness, even if not generally recognized, to human perseverance, ambition, and hardiness' (Caillois 1961: 138). Vertigo negates instrumental benefit and embraces risk for its own sake and the affirmation of human bodily experience. By allowing for the arousal, exposure and satiation of forbidden desires, it helps people to more fully be themselves.

Play as vertigo most clearly illustrates a general emphasis within this book upon the more active forms of playful engagement with the built environment. Many different kinds of people use the city for play, and focusing on the more kinetic, risky and transgressive forms of play tends to also draw attention to the practices of those social groups who have the capacity and disposition to pursue them, in particular young men, a group who have the enthusiasm and the time to explore new-found bodily potential. It is in many cases from the play of 'able-bodied' young men that inferences can be drawn about how people in general might explore the physical and social limits of urban spaces.

There are many ways in which the city frames disturbing sensations which stimulate vertigo. The following extract is from Caillois' analysis of funfairs. Rereading this account as a description of cities highlights many ways that experience in urban space can inspire vertigo.

These physical sensations [of vertigo, from rides] are reinforced by many related forms of fascination designed to disorient, mislead and stimulate confusion . . . this is the function of labyrinths of mirrors (. . . the disconcerting reflections that multiply and distort . . .) and of freak shows exhibiting giants, dwarfs . . . creatures that are . . . half-woman and half-octopus, men whose skins have dark spots like those of leopards. The horror is compounded by being invited to touch them. Facing these attractions are the no less ambiguous seductions of phantom trains and gloomy haunted houses filled with apparitions, skeletons, entangling spider webs, bat's wings, trap doors, drafts, unearthly cries . . . a naïve arsenal or miscellany of terror, adequate to exacerbate nervousness grown complacent and generate a fleeting horror . . . everything that is strange or disturbing is of use here.

(Caillois 1961: 134–35)

Caillois' litany of phenomena suggests the sheer intensity of sensory stimulation in the city. Urban space is physically extreme, consisting in part of giant forms and dark and compressed spaces which contribute to a sense of the sublime, 'the awe-filled pleasure of submission to that which overwhelms us – a mixture of reverence, fear, and an almost phallic pleasure inspired by grandeur' (Dovey 1999: 120). Such mixture characterizes the 'ambiguous seductions' which the city presents. The sublime also implies an inability to rationally apprehend. In the city, perception is constantly desta-bilized by sudden shifts in scale, views of tall buildings and rapidly moving vehicles as well as views from them, and contrasts of light and darkness and of noise and quiet. Unknown spaces suddenly become revealed; similarly, familiar routes, places and meanings become obscured and lost. The city is disordered, a 'labyrinth of mirrors that multiply and distort'. Benjamin and the Situationists suggest that wandering through the labyrinthine spaces of the city brings perceptions of people, places and memories together in strange, random juxtapositions, dislocated from their conventional frames of reference:

> walking . . . is a substitute for the legends that used to open up space to something different . . . Things extra and other (details and excesses coming from elsewhere) insert themselves into the accepted framework, the imposed order.
>
> (de Certeau 1993: 160)

The social and physical complexity of urban public space frames disturbing 'dialectical images' like the half-woman-half-octopus, images which shock the senses and also one's sensibility about the order of things (Gilloch 1996).

The fear of touching freaks which Caillois mentions generates moral vertigo. In urban public space, a great many people who are strange and different are gathered close together and encountered by chance. A myriad of social activities overlap and interpenetrate (Sennett 1971). Caillois writes of 'the pleasure, thrill and excitement engendered by fraternization with an anonymous multitude' and adds that 'collective turbulence stimulates, and is in turn stimulated by, [vertigo]' (Caillois 1961: 40). Thus play in public also feeds upon the disorderly behavior of crowds of other people.

Vertigo 'is a game-characteristic which Caillois appears to have "discov-ered" and which has received little attention from others' (Lancy and Tindall 1977: 12). The conception of play as vertigo has strong resonances with urban conditions. This is particularly clear where Gilloch points to the origins of Benjamin's phenomenology in Baudelaire, who saw the city as:

> a site of intoxication . . . home to the unexpected, to novelty and distraction . . . a space to be explored with joyous abandon. It offers

the excitement of the anonymous crowd, the exhilaration of freedom and the ecstasy of losing oneself. It is a place of shock.

(Gilloch 1996: 171–72)

Vertigo and simulation are closely linked and often occur together. This can be seen in many kinds of performance and costumes: the alternative world created through play draws on memory and dreams, but it also distorts and intensifies; it is both strange and larger than life. People often combine disguising themselves to assume another personality with special kinds of movements, such as dancing, which not only lend them another character but also transform the way they experience the world around them. Amusement parks are play spaces which draw together themed environments and the thrill of rides. Play as simulation and play as vertigo are shaped in response to the atmosphere of the city and the tension generated by its coincident familiarity and unfamiliarity.

Caillois' four categories have some claim to being a comprehensive delineation of play behaviors (Mouledoux 1977). Each is a different way in which life is lived more intensely, and each suggests forms of heightened bodily and mental engagement with the rich specificity and strangeness of urban space. Competition and simulation focus on increased personal control over the body and over communicated meaning. Chance and vertigo involve escape from behavioral and perceptual controls. In each case, play is experienced as an emancipation from the routines, constraints and preconceptions of everyday social existence. While competition and chance relate primarily to material relations between people and urban spaces, simulation and vertigo are forms of activity which primarily reflect Benjamin's interest in play as a transformed mode of perception in the city.

Competition and simulation, the two more ludic forms of play, clearly have an analogic role in inculcating and reproducing social *habitus*, both for children and for adults. However, Caillois notes that playing with chance depends on an adult's capacity for rational foresight, and that vertigo lacks any educative value. The emphasis by psychologists such as Piaget on the developmental function of cognition fails to explain why adults remain attracted to chance and vertigo – reinforcing the fact that different scholars use the term 'play' differently to convey and explore disparate aspects of human experience (Mouledoux 1977; Sutton-Smith 1997). Chance and vertigo suggest something other than rational instrumentality as a driver of social behavior.

Caillois' four forms of play help to elucidate the dialecticity of play, because each of these forms expresses social ideals which are difficult to achieve in everyday life. Equalization through competition or chance emphasizes a negation of predetermined, unequal social roles. Simulation and vertigo occur in an improvised world which is without fixed rules; players either create an illusion or are dominated by it. The focus in simulation and vertigo is on liberation from power relations, from typical methods and from normalizing constraints, whether these be psychological, social or natural. Freedom from predetermination, a certain improvisation of practice, is a basic condition of play. In public places, people can fully 'be themselves' and can transcend the roles which have been defined for them by work and domestic life. The four forms of play act out and amplify the potentials for freedom, equality, individual control and transformation which are only latent in the diversity, intensity and reduced social structuring of urban public space.

There does not appear to be any previous research that specifically examines the spatial conditions that frame Caillois' various types of play. However, each of the types would seem to be suited to very different kinds of physical settings. Urban spaces can heighten people's control over their own bodily action and their presentations of self. Urban spaces can allow people to slip out of the modes of social and bodily control which normally govern everyday life. And urban space clearly frames unexpected experiences and unplanned encounters with strangers.

The publicness of play

Play in urban public settings has a distinctive phenomenology and sociology. Caillois' four types of play highlight that play activities define special ways of perceiving and certain modes of interacting with other people, objects and spaces. Play is shaped by urban social conditions: the density and diversity of people, the mixing of their activities, the unpredictability of their behavior, their differing expectations and the unfamiliarity of their expressions all contribute to instability and 'the dissolution of constraints' (Lefebvre 1996: 129). Lofland (1998: 77–98) places the pleasures of public life into two main categories, aesthetic and interactional. Her list of aesthetic pleasures includes the experience of historical layering and physical juxtaposition, and the diversity of stimuli this frames, very much paralleling Benjamin's focus on urban perceptions. Lofland also lists unexpectedness (chance), crowding (vertigo) and whimsy (simulative fantasy). These aesthetic perceptions are triggers to play through a combination of sensory stimulation and memory. The forms of interactional pleasure which Lofland outlines are public solitude, people watching, public sociability, and lastly what she categorizes as playfulness, frivolity and fantasy. Because of the extensive freedom of movement and action in the city, there are many different levels of involvement which people can take in public play.

Several overarching principles guide and structure public face-to-face interactions so that everyone can cope with the city's social intensity and complexity. The principles of 'civil inattention' (Goffman 1980) and 'civility toward diversity' embody the freedom and social distance to act playfully in the midst of multitudes of strangers: people either ignore or tolerate play. Another principle of public interaction is that 'inhabitants of the public realm act primarily as audience to the activities that surround them' (Lofland 1998: 31). People watching is a pleasure in itself. It allows people to fantasize about the lives of others. This form of play relies on the visibility of strangers in public spaces, and to a lesser degree the viewer's own exposure. Being seen in public is just as important as seeing. The presence of an audience is an important part of many play practices, most notably competitive displays. The reactions of all those present during play activity in some way define and legitimize or invalidate its boundaries, even bystanders who apparently just observe or turn away. 'Games generally attain their goal only when they stimulate an echo of complicity. . . . games . . . seem to reflect stimulus and response . . . and effervescence or shared tension. They need an attentive and sympathetic audience' (Caillois 1961: 39–40). The mere presence of other people raises the stakes of contests. Audiences help judge the fairness of play, and often enforce it, as well as its success. The public are the context of public play and they are an integral part of the activity's meaning (Lutfiyya 1987).

Play events in public can encourage bystanders to join in, taking on a more active level of public engagement, or they can inspire related events. The most distinctive feature of interactions in the public realm is that the majority of people are biographically unknown to each other. What urbanites know about each other is generally only categorical, limited to such characteristics as their age, race and gender. Urban play thus primarily involves relatively impersonal interactions among strangers. The diversity, anonymity and unfamiliarity of other persons encountered in the city lend public play a distinctive character (Lofland 1998). The limited scope of knowledge and involvement allows people to escape from their normal social status and responsibilities with a low level of exposure risk. As people in public do not always know enough about each other's status, capacities and motivations to know the 'appropriate' way to interact, play provides a relatively low-risk way to test the boundaries of the other. People's playful responses constantly test, dissolve and invert established behavioral cues, strategies and meanings. The presence of others encourages and shapes this play. Other people can be a source of wonderment and fantasy; they can be the basis for competition and simulation and engender the collective turbulence of vertigo. 'Much behavior in public is not confined to specific tasks': in public play, the common purpose is merely to be with and to experience other people as ends in themselves (Lennard and Lennard 1984: 9).

Play is often interactive, especially in public places. The separateness of play from people's everyday life heightens their awareness of others who are participating with them. Freedom from social responsibilities allows for the broadest expression of people's individuality. Others can experience different aspects of a person's personality through their play, and it then becomes less likely that their personality will be conflated with the specific social roles they fill in everyday life, and less likely that they will be treated instrumentally (Lennard and Lennard 1984). Interactive play in public places can help to build feelings of connectedness and community; it can draw heterogeneous people together. It is 'pure' sociability which is not distorted by conflicting individual goals (Simmel 1950). Public play relations do not have a thoroughly predetermined structure. Play's voluntary nature means people are free to interact on openly negotiated and hence more equitable terms. Informal and chance interactions can build over time into more permanent relationships.

Lutfiyya (1987) articulates three hierarchical levels of play at different scales of the social: individual play which develops cognition and the imagination, social play which develops solidarity within social groups, and public play. For Lutfiyya, the distinctive characteristic of public play is the extent and impact of the players' acts of 'fanciful recontextualization'. Fanciful recontextualization means the critical redefinition of serious social contexts, and so provides a link to the general dialectical character of play.

The dialecticity of play in the public realm

Play can be understood within Bourdieu's (1984) framework of *habitus* as a practice which is 'structured' and also 'structuring'. Acts of play arise within a cultural context and help to reproduce it, and play can thus become 'an instrument of fecund and decisive culture', serving material and ideological outcomes and becoming 'diffused to reality' (Caillois 1961: 27, 64). Both Caillois and Huizinga compare play to a range of ritual social forms including art, war, poetry, myth, philosophy and religion; both see the whole of culture as played. Play can also impart new meaning and potential to society: 'Play form emerges from the contents of ordinary or serious life situations, but ultimately is not bound in these contents. Play as a transformational process develops an "autonomous existence"' (Lutfiyya 1987: 10; see also Simmel 1950). But rather than treating play as a completely separate category of human experience, it is more useful to use it as an analytical construct for understanding how everyday life unfolds dialectically.

One of the main defining characteristics of play is its tension with everyday life: 'Play forms typically involve testing . . . Through play *homo ludens* lives out emotions which are either repressed or diverted by the rest of life' (Rojek 1995: 185). The four-part definition of play laid out in this chapter points toward several oppositional tensions between play and various 'serious' features of everyday social life. These include:

morality	desire
order	disorder
intention	surprise
production	destruction
deferred satisfaction	immediate pleasure
control	release

Play provides opportunities for critiquing, transforming and expanding social practice because of its diversity and creativity, its testing of limits, the absence of instrumental gain and its separation from the roles, rules and expectations of everyday life. Two further aspects of the dialecticity of play pertain to the particular relations between social action and physical space. The first is that the built environment in general is a particularly durable part of *habitus* which inspires and gives structure to the actions of everyday life. Built forms tend to suggest what behavior is 'appropriate' or 'desirable':

> the understandings that prevail among all socialized persons present at the occasion, the territorial plan, the physical objects present, and the social usage of any or all of those objects and spaces constitute criteria for understanding what is going on, what is supposed to happen.
> (Lyman and Scott 1975: 149–50)

Spatial *habitus* constrains play because it defines 'appropriate' times and places which help to organize and codify social action, and because it has representational contents that reinforce social norms. However, built space also 'contains potentialities – of works and of reappropriation . . . responding to the demands of a body' (Lefebvre 1991b: 349). Space offers certain points of escape and resistance. New forms of social practices arise in critical response to the conditions of space. Practices are spatial in a dialectical sense; they also produce space: 'a body . . . putting up resistance inaugurates the project of a different space' (Lefebvre 1991b: 349).

Social experience in urban public space has its own specific tensions. The *habitus* of urban life is not neatly structured. Urban space confounds expectations; it lacks clear order, it is constantly being used in a variety of ways by others (not all of which appear logical), and it lends itself to new forms of behavior. Exposure in urban public space illuminates contradictions within social life, because it exposes people to the concentrated diversity of the *oeuvre*. In doing so, it also heightens those tensions, and brings about ruptures, where people's experience of life is transformed. When people have unexpected encounters with others who are different in public places, they typically have to actively negotiate their engagement, because they cannot follow predetermined rules of conduct. Sennett (1994) suggests that the liberty of public life is defined not by an absence of constraint, but by this active, heightened engagement with possibility and its inherent difficulties: 'Freedom which arouses the body does so by accepting impurity, difficulty, and obstruction as part of the very experience of liberty . . . The body comes to life when coping with difficulty' (Sennett 1994: 9–10).

A multiplicity of practices of play, discovery, and negotiation continuously re-shape everyday urban life, through a dialectical engagement with spatial and social discipline which undermines the possibility of order:

urban life increasingly permits the reemergence of the element that the urbanistic project excluded ... the city is left prey to contradictory movements that counterbalance and combine themselves outside the reach of panoptic power ... they are impossible to administer ... a proliferating illegitimacy ... discipline is manipulated ... spatial practices in fact secretly structure the determining conditions of social life ... multiform, resistant, tricky and stubborn procedures that elude discipline without being outside the field in which it is exercised ... The surface of this order is everywhere punched and torn open by ellipses, drifts, and leaks of meaning.

(de Certeau 1993: 155–57)

The social experiences framed by urban space can reveal, and indeed stimulate, new needs and new possibilities; they create, reawaken and resituate meanings. Lefebvre's (1996) formulation of the dialecticity of everyday life is captured in his conception of 'moments'. Moments are temporally limited social experiences, characterized by conditions under which the oppositions and contradictions of social life are intensified, thereby raised to consciousness, and engaged. Lefebvre's idea of play as a moment is thus far more dialectical and creative than Huizinga's formulation of play as occurring separated from the everyday:

A moment defines a form and is defined by one. The everyday is composed of a multiplicity of moments, such as games, love, work, rest, struggle, knowledge, poetry and justice, and links professional life, direct social life, leisure and culture ... when playing, one accepts the rules of the game and each time recreates and reinvents the usage of the game.

(Kofman and Lebas 1996: 30)

In their capacity as transformative moments within the everyday, acts of play are akin to a broader class of dialectical social practices, rituals. Anthropological knowledge about the ritual process, which is itself often defined in terms of play, can shed much light on play's causes and outcomes, how it is conducted and experienced, and its social and spatial context. Many social rituals unfold in time in a process which Turner (1982) describes as being *liminal*, from the Latin *limen* for threshold. The ritual process is, like play, a dialectical moment:

an interval, however brief, of margin or limen, when the past is momentarily negated, suspended or abrogated, and the future has not yet begun, an instant of pure potentiality when everything, as it were, trembles in the balance.

(Turner 1982: 44)

The concept of liminality charts the unfolding tension between the passage of a social ritual and enduring cultural definitions of rationality and value. Liminality, like play, frames escape from social convention and the exploration of new possibilities. The dissolution of social order within the liminal process is always temporary: 'forms of reversal . . . occur during interstices between periods of intense or serious activity' (P. Stevens 1991: 238).

The first phase of the ritual process is separation. The ritual has an 'antistructure': it 'inverts or dissolves the normal (and normative) structural order prevalent in the rest of the community', destroying the truth and power of this order (Spariosu 1997: 33). The bonds of the past are negated. There is the active transgression of rules, an experience of escape or release. In acts of play this is often manifested in the risking of limits and or exposure to intensity and excess. At the same time, liminality also blurs distinctions: it 'brings together dichotomous elements of life' and '[combines] the ordinarily uncombinable' (Lyman and Scott 1975: 151). Play admits the possibility of bridging the binary oppositions which normally define social life, and engaging dialectically with the tensions inherent in them.

In the second, transitional phase of ritual participants are freed from their usual roles and status relations and from behavioral norms. Other people, symbols and objects are all encountered outside cultural frames of reference and ordinary instrumental relations. The dissolution of norms leads to a leveling of status, as well as ambiguity and an ambivalence regarding purpose; that is, a disinterestedness. People go through 'an ambivalent social phase of limbo' (Spariosu 1997: 33). Goals are unknown and undefined. People rediscover a reciprocal relation to objects and to each other as all things are set free of their cultural coordinates and once again become auratic: very much as in child's play (Gilloch 1996). The transitional phase of ritual is one of heightened awareness. Experience and gesture are sensualized and enriched. The heightening of perceptions and the need for personal orientation stimulate both memory and fantasy. Similarly in play: 'Participants often experience a heightened awareness of others and their surroundings, and are reminded of their connectedness and continuity with the past' (Lennard and Lennard 1984: 53).

The unstructured and unfamiliar conditions of the transitional phase encourage a spirit of inquiry, ingenuity and flexibility. Ritual liminality is crucial to identity formation because it encourages the discovery and development of new understandings of the self through performance. Similarly, 'play is like education of the body, character, or mind, without the goal's being predetermined' (Caillois 1961: 167). People seek to explore options, to establish new meanings and new correspondences between things through metaphorical reinterpretation. Lutfiyya (1987) calls this exploration 'fanciful recontexualization'; Benjamin uses the term 'playful reconstruction' (Gilloch 1996). Radley's (1993) formulation is that liminal acts of ritual or play 'refract' aspects of society rather than merely reflecting them.

In the final, incorporation phase of ritual, the initiate is reabsorbed into the social whole, and so are the experiences and values born of their transformation. Turner argues that forms of social life developed amidst the liminality of play become incorporated into the social structure: 'Turner in effect sees liminality as a game of disorder out of which new orders emerge. He defines liminal situations as "seeds of cultural creativity" that generate new models, symbols, and paradigms' (Spariosu 1997: 33).

Certain actors exemplify liminality, including children, teenagers, elderly people and those getting married. Teenagers are in a transitional phase of life where they try out new kinds of social roles and test limits, and they have a lot of free time. Strangers and new spaces that they encounter provide a measure for their explorations of identity (Sennett 1971). Teenagers respond to opportunities for play which the city presents, but they also tend to create opportunities by working against the established order of the world. Male teenagers in particular engage in play which involves social confrontation, exhibitionism, and displays of bravado.

There are also recognized liminal times including rush hour and smoking breaks, and special occasions such as public holidays and celebrations. Goffman (1980: 21) observes that the decorum of serious everyday life 'is typically subverted momentarily by parades, convention antics, marriage and funeral processions, ambulances, and fire trucks . . . for a brief time'. Many of these ruptures are themselves serious, but all momentarily tear participants and observers away from their normal role responsibilities, suddenly bringing them to focused attention and a heightened state of action. Liminal actions at these times and by these people are generally publicly tolerated because they are not seen as lasting or threatening; indeed, these liminal stages of release and transformation are necessary for the stable reproduction of the wider society (Bakhtin 1984).

Play too can become formalized through rules and bound up in other more instrumental patterns of practice; the path from *paidia* toward *ludus* can be viewed as a continuing process of institutionalization which forms the basis of cultural life and cultural progress (Caillois 1961). In contemporary secular society, liminal experience has become closely associated with everyday leisure (Cohen and Taylor 1978; Rojek 1995). Since the late 1970s, a large amount of public space in western cities has become carefully restructured and managed to provide leisure landscapes with the appearance of permanent festivity (Hannigan 1998; Dovey 1999). The city's *oeuvre* and the sense of escape into 'pure sociability' have been captured to serve the interests of power and profit, coupled with continued efforts to suppress play's dialecticity, its potential for social transformation.

However at the heart of leisure activity remains 'a constant vacillation between tension and release'; even socially designated 'liminal zones', where transgressive behavior has become legitimized, 'can never be areas of either genuine freedom or genuine control' (Rojek 1995: 87–88). Even though the ritual conduct of play, with its sense of release, its negations, and its apparent risks, might ultimately serve a progression 'from turbulence to rules' (Caillois 1961: 27), what remains of interest is that conditions of turbulence and desires for disorder persist, and that the most complex of human creations, the city, provides a wealth of these conditions. Liminality is an intrinsic characteristic of urban social experience, if not always recognized as such. The city's sensory intensity and unfamiliarity, the unexpected juxtapositions of people's activities in time and space and overlaps of meaning all help to constitute liminality. People's encounters with difference and the unexpected in public space are escapes from the everyday which continue to transform their sense of self (Cohen and Taylor 1978). The liminal, transformative potential of play in urban public spaces cannot easily be suppressed because it resides within people's everyday bodily experiences. Everyday actions have a role in the continual structuring of the social world, developing people's understanding of who they are and who they want to be and their understanding of how they relate to spaces and to other people around them, and expanding their capacities to act.

Chapter 3

The social dimensions
of urban space

The mechanics of play are a product of people's perceptive and performative abilities and of the specific physical contexts where they act and interact. It is easy enough to think, speak and write of urban space in the abstract, claiming that it is a site of freedom and believing that urban life is characteristically and distinctively diverse and intense and full of novelty. These concepts attain concreteness when they are produced through people's bodily experiences of other people and other people's activities within actual spaces (Lefebvre 1991b).

This chapter explores the spatial relations between people when they play. Most play in urban spaces occurs in the presence of other people. Spatial relations are in themselves a significant aspect of how certain types of play take place, particularly competition, chance and simulation. The possibilities of playful interactions among strangers are shaped by the distance and orientation between bodies, and the postures and gestures through which people shape their social encounters: what Hall (1966) calls 'proxemics'.

Interpersonal distance

Physical space plays a significant role in social interactions, because people's lived experience of strangers comes to them through the senses. The distance at which other people are encountered affects how aware people are of each other's characters, moods and intentions. These distances also determine people's bodily capacity to act in response to such stimuli. For these reasons, interpersonal distances come to be understood as representing particular social expectations of affinity and conduct.

Hall (1966) defines four distinct scales of spacing between people: intimate, personal, social and public. These scales are determined by the kinds of sensory information people can transmit and perceive about each other and the kinds of physical interactions they can undertake, and hence the sociocultural expectations which are framed by their separation (Tuan 1977). At each of four interpersonal distances, bodies are able to have a particular range of experiences which people recognize as distinctive kinds of social encounters.

The most distinctive feature of *intimate* distances, less than 0.5m separation, is that people can easily reach out and touch any part of each other.
Strangers generally try to remain outside this distance because 'it is taboo
to relax and enjoy bodily contact with strangers' (Hall 1966: 111–12). One
reason that public spaces in the United States are large and open is because
Americans feel crowded if they have to move within the private space of
someone they do not know intimately (Rodaway 1994).

Certain kinds of person-to-person communication operate only over these
short distances. Smell and touch are intimate senses because they operate
effectively only within the reach of the body (Rodaway 1994). Whether
talking or not, people who are closer together are sharing more detailed
information about themselves and their emotional state. For example, in
crowded situations, even when people avert their gaze, stiffen and restrict
their body movements to maintain a sense of social distance, their ability to
smell others inevitably increases, and this is itself a greater level of
involvement (Sommer 1969). Two forms of haptic sense data or 'body
knowledge' are important in close encounters. First, there is thermal
perception. This is particularly relevant to how a person experiences
crowding: the cause of discomfort is partly thermal, 'when there is not enough
space to dissipate the heat of a crowd' (Hall 1966: 54). Perception of body
heat affects how involved people feel with others, as suggested by such figures
of speech as 'a heated argument' and 'he warmed to me'. Second, there is
tactility, direct pressure on the skin. Third is kinaesthesia, which is 'the ability
of the body to perceive its own motion . . . the movement of the body parts
and the locomotion of the whole body through the environment' (Rodaway
1994: 42). This relates to the experience of vertigo, which is the body's
experience of extremes of scale and motion in relation to its environment,
and failing to make sense of these perceptions.

Relations of touch are often seen as profane, because they escape
socialized frames of reference and reverence. At least five different aspects
of the sense of touch indicate its importance in playful relations among people
and between people and built space (Latham 1999). Each of these can be
most easily illustrated by comparing touch with the dominant, most auratic
sense, vision.

The first difference between sight and touch lies in the way people make
sense of their perceptions. Vision tends to comprehend the world as complete,
fixed images. Gazing on someone objectifies them, and the same is true for
places. As one individual approaches another in public space, they can see
them in increasing detail, and preformulate attitudes toward the image that
they project; people operate on preconceptions. Vision is tied to memory
and *habitus* because people gather these perceptions into a 'world' (Lefebvre
1996). By contrast, tactile sensations remain fleeting and partial. The proximity of urban life means that tactile sensations constantly assail the exposed
body. In public open spaces, these influences are difficult to control; people

are to a certain degree abandoned to random and unfamiliar tactile encounters with the environment and with other people (Goffman 1982). Therefore the action of touching is inherently spontaneous, strange and incomplete, and lends itself to continual exploration.

A second quality of touch is its distracted nature: through touch, the body 'feels its way around the place it finds itself rather than fixing that place with a distancing look' (Latham 1999: 463). Practices of the gaze serve to keep others at a distance. Touch, however, is both spatially and temporally immediate. It cannot be duplicated or transmitted. Touch thus tends to avoid the abstraction, reproduction and accordant social codification which befall images (Lefebvre 1991b). The richness and intensity of tactile relations accentuate the specificity of the person or thing encountered. Touch renders things in greater three-dimensional detail, giving a richer understanding of their mass and texture, how they are structured and how they work. The closeness and tangibility of tactile experience with other people emphasizes interactive relations as opposed to passive perception.

Third, vision contrasts with touch because of vision's tendency toward instrumental function. The sense of touch serves practical purposes, but while someone is preoccupied performing tasks, it constantly perceives, in a state of distraction or inattention. Through touch the body remains open to sensations that serve no practical purpose, and this provides an avenue for the stimulation of non-productive desires, through the awakening of memories, calling upon habit, or spontaneous reaction.

Fourth, because tactility relies on physical contact, it is a sense with a high degree of reciprocity. Touch is an 'embodied' and 'corporeal' way of knowing others (Latham 1999). One's own body must be present to touch, it is not possible to remain distant. Touching someone is also always a direct stimulus upon one's own body: to touch is to be touched (Rodaway 1994). When people brush up against strangers in public space, they cannot help but perceive them and react. Touch is a particularly rich dimension of social encounters, because one learns about one's own body as well as about others.

The sense of touch thus emphasizes the equality and fragile tangibility of all bodies in space. Close distance and touching embody greater risk and require a greater level of trust (Sennett 1994). This relates to a fifth observation: despite claims as to its power in western society (Foucault 1977, 1997), vision is passive. Touch is not just a form of perception, but is always a form of action which has the capacity to directly transform the space or the person perceived (Rodaway 1994). These dimensions of the sense of touch illustrate why it is the most intense form of encounter with strangers and the built environment. The immediacy, reciprocity, non-functionality and transformative potential of touch suggest ways that bodily contact frames opportunities to explore playful desires.

People's sensations beyond the intimate distance, at *personal* distances of 0.5 to 0.8m, still involve a kinesthetic sense of closeness. People remain bodily aware of the possibility of touching or holding the other person. At this distance, the 'three-dimensional quality of objects is particularly pronounced' (Hall 1966: 113). Between 0.8 and 1.2m, the other person is kept at arm's length. Physical contact relies on mutual agreement: two people can touch hands only if both reach out. Thus this is 'the limit of physical domination in the very real sense. Beyond it, a person cannot easily "get his hands on" someone else' (Hall 1966: 113). Engagement is personal and exclusive, yet each individual also maintains a certain privacy. This is a common distance for casual interpersonal conversation in public settings in North American and Northern European cultures (Fast 1971). Whyte (1980) found that the 'effective capacity' of the seating in the busiest outdoor plazas in New York City was between 108 and 125 people for every 100m of the available length of sitting space. This confirms that by choice people stay at arm's length (0.8m), and are unlikely to sit within a close personal distance of strangers, even when the desire to sit is great and options are limited.

Social distance ranges from 1.2m to 3.6m. At 'near' social distances of 1.2m to 2.1m, nobody touches or expects to touch unless there is special effort. Intimate visual detail in the face is not perceived, and other cues to people's disposition such as body heat and odor are not generally detected. This 'is a very common distance for people who are attending a casual social gathering' because 'at this distance a couple can engage each other briefly and disengage at will' (Hall 1966: 115–16). In informal public leisure settings, strangers often stay within such a distance, where contact remains a possibility. At large open spaces such as beaches, for example, people do not spread themselves evenly over the entire available area; rather, 'they go where the other people are ... in checkerboard fashion, they located themselves about one or two spaces removed from the other people' (Whyte 1980: 68). A couple lying down occupies approximately 2m × 2m, and hence 'one or two spaces' implies a 2m to 4m social distance (Scheflen 1976).

Based on detailed participant observation of social interaction among strangers in bars, Cavan (1966) finds that distinct areas within a typical bar layout facilitate interactions among strangers at a personal distance while limiting the risks. Forced proximity between strangers near the physical bar makes unplanned encounters more likely there. People adjust their behavior and their social tolerance under such crowded conditions: 'by reducing the distance between patrons, both the propriety and the practical problems of making contact are minimized' (Cavan 1966: 104). In the 'milling area' of bars, patrons are able to easily initiate interactions with many strangers. Constant small movements change people's level of exposure to others, either accidentally as they shift their stance, or surreptitiously as they inch toward or away from someone, without such moves necessarily being obvious, well-defined or decisive. Interactions can be adjusted at will by any party, face can easily be saved if necessary, and hence there is less risk involved in any interaction with a stranger. Cavan also finds that three bar stools (a distance of 2.4m to 3.0m) is the maximum separation over which people initiate encounters. In the bar stool setting, if the other person is successfully engaged, the first party generally attempts to close the gap, to minimize the risk of someone sitting between, as the obstruction of eye contact generally termi- nates encounters. At social distances, maintaining eye contact is a particularly important way of communicating continued involvement with the other party, because vision becomes the only form of exclusive engagement (Fast 1971). It is also a means by which an individual maintains their dignity and communicates their expectations when a space that they have to themselves is invaded by a stranger (Sommer 1969). Three meters is too far removed to presume familiarity and the possibility of physical contact, yet engage- ments can be initiated because each individual can read the signals of temperament projected by the other. The encounters Cavan observes on bar stools move across the boundary from a public to a non-threatening private engagement.

At 'far' social distances of 2.1m to 3.6m, social discourse begins to take on a formal character. Interactions can still be initiated at such distances because it is possible to gain and hold someone's attention in a relatively direct and private manner. At distances of 2.4m to 6.1m, a person's actions and comments communicate to a group and not to an individual, at least in the United States (Hall 1973). This distance is thus ideal for performances in the mode of competition and simulation, where visibility to an audience is important. The range of social distance between 1.2m and 3m appears to be critical for playful social relations between strangers because it allows the balancing of safety and personal control over one's level of involvement. This is important in the context of Huizinga's (1970) and Simmel's (1950) arguments that freely chosen and equal participation is an essential component of playful interactions among strangers.

Hall describes *public* distances above 3.6m as 'outside the circle of involvement' (Hall 1966: 116). People at this distance are not necessarily engaged with each other at all. Approaching pedestrians make brief eye contact when they are 6m apart in order to determine their passing direction. As they walk closer, they then avert their gaze so that there is no expectation of interaction, and each can maintain privacy as they pass within each other's personal space (Goffman 1980; Whyte 1988). People initiate engagements with friends once they have approached within the distance of 3.6m. They orient their face and body toward the other person and make facial displays of recognition. As they start to move toward each other, they exchange salutations, and as they come closer they extend hands to shake, or move into position for a kiss (Scheflen 1972).

At distances greater than 4.9m, the bodies of other people viewed begin to lose their roundness and look flat. Two-way communication by hearing is efficient only up to about 6.1m, and even then sounds tend to become more ambiguous with distance (Hall 1966). Above 5m, sensory information about other persons, other than appearance, becomes diminished and unreliable. At such distances, people are effectively flat images to be gazed upon and read, rather than solid, sensual bodies to be encountered and negotiated.

Beyond the sphere within which people can be engaged in two-way communication, sight continues to provide much information which shapes an individual's developing course of action. Over a distance of about 25m it is possible to determine the mood and feelings of others, and this is among the reasons that the maximum distance between a theatre stage and the audience is around 30m, a scale within which 'primary feelings are communicated' (Gehl 1987: 67). Alexander et al. (1977) suggest about 18m is an ideal diameter for neighborhood public squares. His justification is twofold. First, the maximum distances across such a square allow people to recognize people's faces and to half-hear conversations. Second, he assumes an average distribution of 28m^2 per person, which is equivalent to a personal space of 3m radius per person: ideal for informal, unforced social encounter among strangers. This rule of thumb means that a plaza with a diameter of 18m would require only twelve occupants to seem lively, but would still present a significant diversity of potential social engagements. Larger plazas need more people to seem lively and are thus suited only to the busiest parts of cities, although inevitably they more often seem deserted and always feel more anonymous; 100m effectively circumscribes the maximum 'social field of vision', because this is the distance at which one can still recognize people by such details as how they walk, their clothing and their approximate age (Gehl 1987). This is not to suggest large plazas can never inspire playful behavior; merely to note it is at specific smaller scales, determined by the body's own perceptions and movements, that the urban landscape plays a role in shaping a user's encounters with other people and objects.

Two distances seem particularly significant to framing play in public spaces: 25m is an approximate maximum distance for developing an adequate recognition of what is going on in a public social setting. Hall's (1966) 'far social distance' from 3.6m down to 2.1m is a crucial interpersonal separation, the distance at which strangers first acknowledge each other. This is an important threshold at which people consensually determine their mode of encounter. It is a distance at which people metacommunicate that it is acceptable to make overtures. Within this distance, bodily interaction begins to take precedence over the gaze.

Interpersonal orientation

A second explanation Cavan (1966) gives for the frequency of stranger interaction at bar stools is their arrangement. In the 'lateral' (side-by-side) position, people's personal boundaries are less clear, and thus they can be crossed without necessarily constituting a transgression. Few social expectations are framed in a lateral arrangement of bar stools. They are neither sociofugal, which 'tend to keep people apart', nor sociopetal, which 'tend to bring people together' (Hall 1966: 101). The social space of bar stools is 'fluid and unstable' (Sommer 1969: 122). Both protagonist and respondent retain a great measure of control over the definition of any given interaction. Face engagements can easily be broken off (Goffman 1980).

Bar stools and other chairs in public settings can also be placed closer side-by-side than opposite each other. Personal space is smaller to the sides and back of a person than it is in front of them, where vision extends their apprehension of those approaching. Seating which orientates people face-to-face with strangers is confrontational, best suited to competitive encounters. Directness of eye contact can be a more significant determinant of people's level of psychological engagement than the distance between them. People generally prefer to converse with strangers seated diagonally across from them, because this is less threatening, and people who are on intimate terms or working together prefer to sit adjacent (Hall 1966; Sommer 1969). People are socially comfortable sitting back-to-back very close to strangers. If outdoor sitting space is limited, they do so even if benches are as little as 600mm deep and they have to perch uncomfortably (Whyte 1988).

Senses other than vision are generally omnidirectional, and are difficult to block or to frame selectively. The concentration of urban life produces an intense array of sounds and smells which can impact unexpectedly and uncontrollably on an individual's chosen course of action. These perceptions are less definitive than vision. They always have an element of the unknown. The fact that vision gives the most focused and detailed perception in large part explains why face-to-face proximity is both the most invasive and the most empowering:

> Frontal space is primarily visual. It is vivid and much larger than the rear space that we can experience only through nonvisual cues. Frontal space is 'illuminated' because it can be seen; back space is 'dark' . . . The front signifies dignity. The human face commands respect, even awe. Lesser beings approach the great with their eyes lowered, avoiding the awesome visage. The rear is profane. Lesser beings hover behind (and in the shadow of) their superiors.
>
> (Tuan 1977: 40)

Close monitoring of a person's face and their activities under the gaze of others is a means to reinforce behavioral norms (Foucault 1977). Thus frontal relations between people are those most strongly framed by cultural conventions as to distance and behavior. These relations also frame social meanings, reinforcing differentiated role relations. People manage their facial expressions and their bodily actions to constitute a 'front':

that part of the individual's performance which regularly functions in
a general and fixed fashion to define the situation for those who observe
the performance . . . around which it is easy for [the observer] to mobilize
his past experience and stereotypical thinking.

(Goffman 1959: 19, 23)

Goffman (1959) argues there is a strong correspondence between the
frontality of social relations in space and the reinforcement of social norms,
describing interactions as often being structured around socially defined
'front' and 'back' regions. Front regions, the social settings where performances are given, frame moral and instrumental demands upon action,
ensuring that actions undertaken there generally serve to reproduce social
structure:

> Performances in front regions typically involve efforts to create and
> sustain the appearance of conformity to normative standards to which
> the actors in question may be indifferent, or even positively hostile, when
> meeting in the back.
>
> (Goffman 1959: 208)

Thus the back region serves a dialectical purpose:

> A back region or backstage may be defined as a place, relative to a given
> performance, where the impression fostered by the performance is
> knowingly contradicted as a matter of course . . . it is here that illusions
> and impressions are openly constructed . . . Here the performer can relax;
> he can drop his front, forgo speaking his lines, and step out of character.
>
> (Goffman 1959: 97–98)

In back regions, 'various potentially compromising features of interaction
are kept absent or hidden' (Goffman 1959: 207). The concept of a back
region is an attempt to formulate the characteristic spatial contexts of sacrilegious, vulgar and undisciplined social practices. The back region parallels
Huizinga's conception of a place apart from seriousness where people are
free and where play can flourish. Goffman identifies the following spatial
relation between front and back regions:

> Very commonly the back region of a performance is located at one end
> of the place where the performance is presented, being cut off from it
> by a partition and guarded passageway. By having the front and back
> regions adjacent in this way, a performer out in front can receive
> backstage assistance while the performance is in progress and can
> interrupt his performance momentarily for brief periods of relaxation.
> In general, of course, the back region will be the place where the

performer can reliably expect that no member of the audience will intrude.

(Goffman 1959: 98)

Goffman suggests there are some kinds of places, such as summer resorts, which 'seem to fix permissiveness regarding front, allowing otherwise conventional people to appear in public streets in costumes they would not ordinarily wear in the presence of strangers' (Goffman 1959: 108). Such locations also allow people a greater range of action. This highlights that the publicness of performance does not necessarily equate with moral and instrumental purpose. Goffman also notes that some regions vary in function on different occasions as either front or back, and in particular, 'a front region . . . often functions as a back region before or after each performance' (1959: 111). This highlights that the definition of a situation as serious or playful can also be established through temporal boundaries. Tuan (1977) briefly reflects upon whether cities are structured with 'front' and 'back' regions. Processional entry routes exemplify the former. The only example Tuan provides of a 'back' region relates to the direction north in traditional Chinese cities, which was the location for profane commercial uses. Giddens (1979: 209) suggests urban slums may be a back region, ' "hidden away' from the time-space paths which others who use the city . . . follow'.

In general terms, informal contacts between strangers are more likely to emerge when their encounters happen in back regions, or through bodily encounters from the side or rear, because such encounters are relatively free of monitoring and structuring:

These conversations can start when people are at ease, in particular when they are occupied with the same thing, *such as standing or sitting side by side*, or while engaging in the same activity together.

(Gehl 1987: 170–71, emphasis added)

Whereas 'contrapositional' (face-to-face) seating defines the other person as the clear focus for interaction, obliging specific forms of either engagement or inattention, lateral seating allows people to be within interaction distance, and to explore and get a feel for the other person, without necessarily maintaining a front. Gehl (1987) notes Goffman's findings that strangers need a reason to enter engagement, and points to Whyte's concept triangulation as a definitively spatial illustration of how strangers can encounter each other in playful experience.

Triangulation is 'that process by which some external stimulus provides a linkage between people and prompts strangers to talk to each other as though they were not' (Whyte 1980: 94). The source of stimulation may be any unusual person, a deliberate performance, music, or an interesting physical object or sight. The quality of the event itself is not necessarily important; 'the real show is usually the audience. Many people will be looking as much at each other as at what's on the stage' (Whyte 1980: 96): again underlining the fact that indirect viewing is more playful than rigidly frontal viewing. What is important is that a triangulating event provides a context where strangers initiate contacts. It does so in three ways. An external stimulus causes people to spend more time in close proximity to strangers, while they are watching. It also directs the gaze and body posture of people in the crowd away from each other, making their close bodily encounters less confrontational. Third, it gives strangers something in common to base an interaction around. In these ways, any novel events in public space can provide a framework for people to observe strangers and then engage with them. Two further aspects of triangulation suggest it can stimulate playful behavior among the audience. First, street entertainment can change people's mood, making members of the audience more relaxed in their interactions with each other. Second, street acts are unexpected. Thus they stimulate interactions between strangers in unexpected times and locations within public space, placing these engagements outside preconceived social roles and serious objectives. This is why these 'moments are true recreation, though rarely thought of as such' (Whyte 1988: 97). The density of these kinds of events and the number of potential onlookers in the city suggest that triangulation is a key characteristic of urban play.

People use a wide variety of body postures and gestures to establish a sense of informality in their social interactions (Scheflen 1972, 1976). They use their bodies to communicate the purpose of their encounter to each other and to onlookers. As they communicate these intentions, they are also defining the social reality of their encounters. Much of the 'surreptitious inching' and 'vague and casual movement' which Cavan (1966) observed in bars is done with the aim of constituting and signaling specific social relations such as trust, confrontation, control and commitment. Such moves can serve the purposes of playful engagement with strangers. People use what Scheflen calls 'stances of affiliation' to indicate they are 'with' other people in a space. These include leaning or moving closer to one another, turning toward one another, uncrossing arms and legs, using body extremities to mark off spatial boundaries and shared foci of orientation, and mirroring of posture or gesture. By contrast, people conversing at a large interpersonal distance, positioned on a 60 degree or 90 degree angle to each other and making eye contacts with other people outside their immediate engagement are communicating that they are 'open' to being approached by others.

The idea that indirect bodily orientation both reflects and achieves indirect psychological engagement highlights the strong links between bodily practice, perception and thought. Things and persons straight ahead of a person, in the line of sight and in the path of forward movement, are typically bound up with instrumental purpose, formality and familiarity (Tuan 1977). Urban experience, however, is seldom so straightforward; it is characterized by a state of distraction which admits great potential for diversion and play. When people's attention and movements are physically directed to matters of focused interest, other directions in space are naturally sectors of relative disinterest which can present opportunities for more playful encounters. Things below, behind and to the side are profane, historically regressive, and 'other' to normative interactions between self and social context. In western culture, the direction left or west has similar connotations of impurity, ambivalence and illegitimacy (Tuan 1977). In urban public space, there are many opportunities for tangential encounters and transgressions of personal space boundaries which are primarily regulated through the forward gaze. It is more difficult to clearly ascertain what objects or people are in one's peripheral vision or behind the body, what their purpose is, and their degree of closeness. Peripheral perceptions are also received in a state of distraction, outside of an instrumental framework. Peripheral encounters with the world are potentially playful because they are less familiar, less precise and less strictly regulated. These relations lend themselves more readily to informal actions of play. One particularly poetic description of the role of peripheral views in framing playful behavior is the 'Gruen transfer', a common practice observed in shopping malls:

> The Gruen Transfer (named after architect Victor Gruen) designates the moment when a 'destination buyer', with a specific purchase in mind, is transformed into an impulse shopper, a crucial point immediately visible in the shift from a determined stride to an erratic and meandering gait.
>
> (Crawford 1992: 14)

A shopper who initially has a purposeful direction of movement and a limited amount of time suddenly senses something else among the richness displayed in their peripheral vision. An image perceived in distraction stimulates ideas and memories. Connections are quickly formed to a multiplicity of desires and possibilities. The shopper's motion then changes to increase opportunities for further such encounters. Walter Benjamin's own *flâneur*-like, distracted, disinterested and tactile way of understanding the city was, in a similar way,

> not sense so much as sensuousness, an embodied and somewhat automatic 'knowledge' that functions like peripheral vision, not studied contemplation, a knowledge that is imageric and sensate rather than ideational.
>
> (Latham 1999: 452)

Peripheral vision does not 'make sense' out of the world. It is, rather, a mode of experience which remains open to the nature of sensation itself, to the formal qualities of things rather than their usefulness or social meanings (Simmel 1950).

The observations of Benjamin and the Situationists suggest that possibilities for play can be framed by new ways of being in, moving through and perceiving space; there is a spatial and behavioral dimension to escapism and creativity. The stroll of Benjamin's *flâneur* and the Situationist *dérive* aim to break down the ordered frontality of the urban scene, its constructed spectacles, its front regions; and to increase the possibility of exposure to more diverse and unusual perceptions and interactions which might be found on tangents, cross-axes, detours and meanders (Plant 1992). Rather than following predetermined paths which reinforce perceptions, the urban walker always constructs their own path and hence actively shapes their own perception: 'walking . . . makes ambiguous the "legible" order which planners give to cities' (During 1993: 151). The complexity and intensity of social action in urban public spaces generates diverse and haphazard sets of trajectories.

While frontality generally implies human purpose, the direction up has cultural connections to sacred ideals and to a future time in which they might be achieved. The placing of a vertical monument on an urban axis is an archetypal instance of these fundamental relations between space and meaning. Conversely, the direction down and the ground itself are linked to the past and to profanity (Tuan 1977). Going below the ground is another way of escaping from the ordered, practical flows of everyday street life and from the normalizing gaze of the public.

People's constant efforts to adjust their relative location, orientation and distance help to create and shape their encounters and, if they so choose, to make them more playful. But they cannot often manage conditions entirely according to their own choosing. Urban life continuously goes on all around them, and can never be completely controlled. The closeness of urban life means an inevitable increase in people's exposure to strangers. Public settings are framed with expectations about certain kinds of sensory exposure, and behavioral norms are adjusted to reflect this. Crowding is made tolerable as it is managed through social practices. Seeing and being seen by strangers is perhaps the most obvious and inevitable form of this sensory exposure. 'Civil inattention' becomes a common response (Goffman 1980). In urban spaces strangers frequently receive information about each other which under normal circumstances is restricted or profane. Such encounters are often random and fleeting. There are social taboos against acting upon them, even acknowledging them.

But the stimulation of dense public life does not lead only to the withdrawal which Simmel (1997) so famously described. Play is also made possible because of the gathering together of a great diversity of people and activities, their adjacency and the varied orientations from which strangers approach each other. Close physical exposure establishes tensions: the management of encounters becomes less a matter of physical limitations and more a matter of social constraints. Urban spatial proximity opens up opportunities for playful transgression of conventional boundaries, as people act upon the knowledge they have of the strangers around them; knowledge which these strangers are not always able to withhold. The physical compression of sensory stimuli means there is much information about strangers which is perceived in a state of distraction, and this stimulates forbidden desires, which may be expressed through play. Proximity enhances the reciprocity of social relations, because it tends to provide heightened sensory information to all parties. Everyone in public has to surrender a certain amount of personal control over time and space to chance conditions. Personal control is also yielded to the broader public's definition of the situation: audiences have a significant role in helping determine whether a certain action or interaction is play (Bateson 1987).

Chapter 4

Paths

The various kinds of paths which make up the pedestrian network in Melbourne's Central Activities District (CAD) offer users varieties of amenity, choice and freedom: opportunities to indulge in unplanned or risky behavior, to linger and to wander. Two major pedestrian paths through Melbourne, Swanston Walk and Southbank Promenade, weave together complex sequences of perceptions and potentials for interactions with other people, and thus provide rich opportunities for play.

The possibilities which urban paths offer for play are not limited to the fixed physical conditions. Play on paths is always given stimulus by the dynamic element of human activity encountered along them. People's creative, playful behavior can transform the potentials framed by the network of paths in the CAD, by introducing new uses for street spaces, new ways of moving along them, and new meanings.

Path type and layout

The major street system of Melbourne's CAD is a uniform 200m grid of four by eight major street blocks, subdivided by four narrower one-way lanes and permeated by a network of smaller lanes and pedestrian arcades (Figure 4.1). Major streets have 30m rights-of-way and are generally four lanes, with tram tracks running in the middle, kerbside parking and wide footpaths. One central block of Bourke Street, between the city's two main department stores, has been pedestrianized and Swanston Walk has been converted into a transit mall running north–south through the length of the CAD.* In between the CAD's major streets runs a series of 10m-wide minor one-way laneways, and an extensive network of ground-level pedestrian arcades and pedestrian-only laneways.

---

* Although the majority of the Melbourne metropolitan area is gridded according to the compass, central Melbourne's grid is tilted approximately 38 degrees west of north. Thus all references to 'north' in the text technically refer to northwest by the compass, and so on. Maps of the CAD grid are shown with N 38° W pointing up the page.

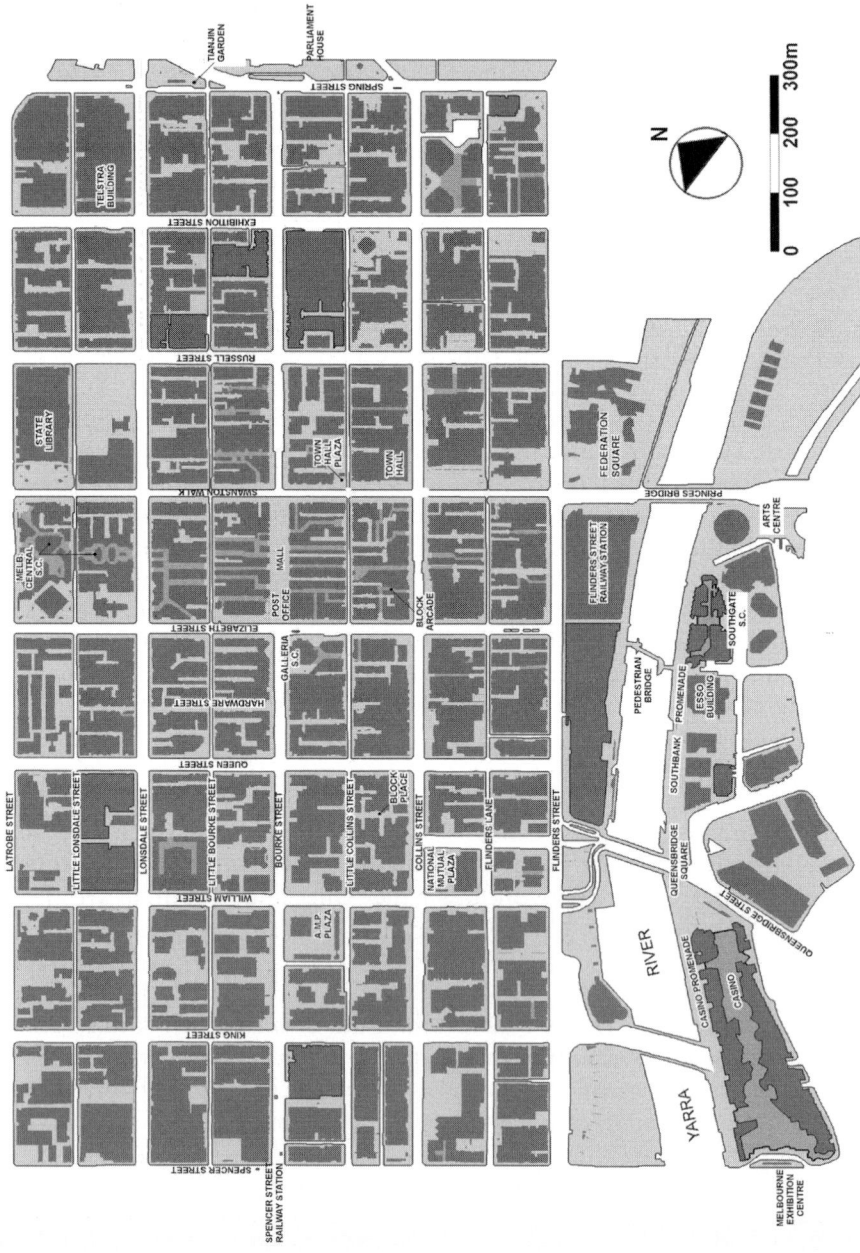

Figure 4.1 Systems of vehicular and pedestrian paths, Melbourne Central Activities District.

Relatively little play occurs along central Melbourne's major vehicular streets, particularly considering that in aggregate they constitute by far the largest area of public open space in the CAD, which has relatively few plazas and only one small park. One possible explanation for the low volume of play on vehicular streets may be that much play occurs during marginal times such as evening rush hours. Pedestrians are presumably disinclined to play near vehicular streets at these times when there is also peak traffic volume and a lot of noise and fumes.

The vast majority of play in Melbourne's public spaces occurs on pathways which are free of vehicular traffic, including vehicular streets when they are temporarily closed. Play which does happen somewhere on major streets most often takes place at their intersections (see Chapter 6).

Some of the playful acts which occur on pedestrian laneways and arcades depend on particular environmental affordances which these street types provide, such as their being little-traveled and protected from weather. For example, at closing time, two young employees of an electronics store drive radio-controlled cars around outside the store on the floor of the Strand shopping arcade. Play on this kind of minor public path is framed by a balance between security and exposure. The enclosure of the space provides physical comfort and the arcade's small scale and low pedestrian volume provides psychological comfort.

The probability of play also appears to be enhanced by greater connectivity and permeability in the circulation network as a whole. Jacobs (1961) argues that urban vitality depends on short blocks, which provide a multiplicity of alternative routes and frequent interconnections between them. When a street system is densely interconnected, any particular street or site which may be a destination in itself is also a secondary or incidental destination within many other orbits of activity (Alexander 1965). A pathway which is chosen by one pedestrian because it is the most practical will also be chosen by others who are seeking playful diversions, and also be used by others who are heading somewhere further afield, increasing the diversity of people and experiences on any given street and the functional integration of various city activities.

The portion of central Melbourne shown in Figure 4.2 highlights sixteen separate minor laneways and pedestrian arcades which cut through the middle of the street blocks between Elizabeth Street and Swanston Walk. These fine-grained pathways provide 168 possible route permutations between Flinders Street Railway Station and the Bourke Street Mall with its major department stores. This system of minor connectors together carries approximately 30,000 pedestrians per day, a comparable volume to each of the major vehicular streets on either side of the figure, which are the busiest in the CAD (City of Melbourne 1985; Gehl and City of Melbourne 1994). These pathways take pedestrians past a rich, fine-grained mix of retail and other uses. Taken in aggregate, three times as many playful incidents occur

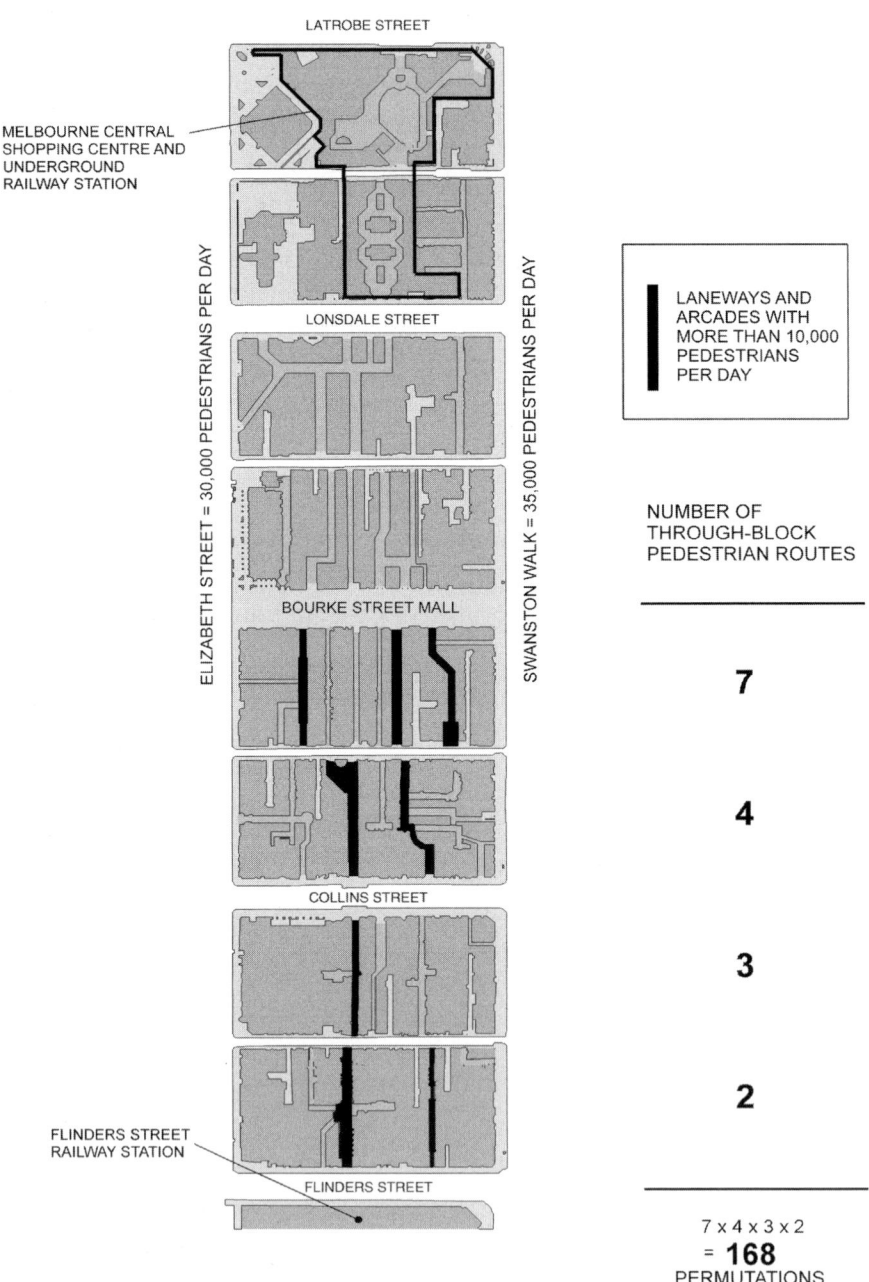

LATROBE STREET

MELBOURNE CENTRAL
SHOPPING CENTRE AND
UNDERGROUND
RAILWAY STATION

ELIZABETH STREET = 30,000 PEDESTRIANS PER DAY

SWANSTON WALK = 35,000 PEDESTRIANS PER DAY

LONSDALE STREET

BOURKE STREET MALL

COLLINS STREET

FLINDERS STREET
RAILWAY STATION

FLINDERS STREET

LANEWAYS AND
ARCADES WITH
MORE THAN 10,000
PEDESTRIANS
PER DAY

NUMBER OF
THROUGH-BLOCK
PEDESTRIAN ROUTES

7

4

3

2

7 x 4 x 3 x 2
= **168**
PERMUTATIONS

Figure 4.2 Alternative pedestrian through-routes between Flinders Street Railway Station
and Bourke Street Mall, Melbourne Central Activities District.

on these through-paths as on the parallel length of Elizabeth Street. More play incidents occur on Swanston Walk, although almost all of these occur where it intersects with minor streets.

The multiplicity of these paths provides more corners and more prospects of people changing direction, enhancing visibility and increasing chance encounters. The minor paths open up the geography of play, providing a range of sites at which play is possible. Having a choice of route when moving across the city is itself a playful opportunity because it can lead to exposure to new phenomena. People can choose to vary their experience. The practice of walking has a significant role in producing each individual's experience of a city (de Certeau 1993); 'walking is more than a form of transport ... Walking also lets people stop, change direction, and experience things' (Gehl and Gemzoe 1996: 51). Using these laneways and arcades, a person can be exposed to a great diversity of experiences during their everyday instrumental journeys across central Melbourne.

The construction of a new pedestrian bridge across to Southbank in 1991 extended the axis of Elizabeth Street across the river, so that together with Swanston Walk it effectively continues the Melbourne CAD's pedestrian-friendly 200m × 200m block structure one block further south. The Victorian Arts Centre, the casino complex and the Exhibition Centre provide the necessary primary attractors which allow Southbank to function much like the dumb-bell of a shopping mall, drawing pedestrians across the river and along the waterfront promenade (Sandercock and Dovey 2002). The extended pedestrian route from Elizabeth Street onto Southbank is unfortunately not as obvious, accessible or commodious as it might have been. A pair of office towers on Southbank provide a landmark visible from Elizabeth Street which signals that the urban axis continues beyond the intervening barrier of Flinders Street Railway Station, but the connecting pathway is squeezed down through a tight underpass which is partly fenced off for the station's subway entry, and at this point the link is barely legible. The likelihood of pedestrians exploring such new pathway connections depends on them becoming part of users' image of the city and being sufficiently easy to navigate (Lynch 1960).

Compared to the CAD itself, the number of different paths a pedestrian can take from the CAD to reach the attractions on Southbank remains small. The former Sandridge Railway bridge, newly opened to pedestrians in early 2006, has increased the choice of routes, but the variety remains limited to the architecture of the bridge spaces themselves and variations in the perspectives which they open up because of their different alignments. Opportunities for further increases in the complexity of this pedestrian network and the functions distributed along it are restricted by the bulk of the railway station, the lack of any inland depth to the Southbank precinct, the unbuildable width of the river and the design constraints of the bridges themselves. Thus the kinds of play which happen along the river bridges and promenade are highly dependent on the design qualities of the routes. Play along Southbank is not supplemented by the unpredictable splitting, mixing and detouring which give the gridded parts of the city its complex, changeable and robust vitality.

Pedestrian-only paths

It is intuitively obvious that people mostly play on pedestrianized streets. A path set aside for pedestrians is also set aside for a whole *oeuvre* of bodily action which is too carefree and risky when one must be on the lookout for cars. The three main pedestrian paths in central Melbourne where people can be observed playing – the Bourke Street Mall, Swanston Walk and Southbank Promenade – provide room for large numbers of people to move. An absence of vehicles also means an absence of noise. Human voices, music and body actions can carry across space and distract passing strangers. While these paths frame environmental conditions suited to play, they also carry the greatest volume of pedestrian traffic, and the amount of play is in general proportion to the number of strangers who are gathered there, and the variety and stimulation that these people provide. Gehl (1987: 27) notes that 'children tend to play more on the streets . . . than in the play areas designed for that purpose', arguing that it is primarily people which attract people to public spaces. Improvements in the amenity of Bourke Street and Swanston Street, including the initial decisions to close them to traffic, have in part resulted from an acknowledgement of their long-standing popularity for pedestrian use. Play has always been a component of the everyday use of these paths. Their closure to traffic has been the result of their popularity for leisure activities, and their closure has then become a further stimulus. In line with significant general increases in pedestrian activity in Melbourne's CAD between 1994 and 2004, weekday pedestrian numbers on these two pedestrianized streets grew by 70 and 25 percent respectively, with weekday evening numbers growing by 200 and 150 percent. The number of people observed engaged in stationary activities on these streets also grew significantly on weekday evenings (City of Melbourne and Gehl Architects 2004). The many people passing time on these streets outside of work hours – as many as 11,000 pedestrians per hour for these two streets alone – clearly shift the balance of street activity from work to play. These two cases illustrate that closing a street to traffic can increase the amount of play there – provided the street is already a popular place to walk and stand around.

Bourke Street Mall is a relatively short section of street space, approximately 200m × 30m. Most pedestrian movements are across this Mall rather than along it, which explains why only this single block was closed to traffic. A majority of its users are stationary: resting on benches, eating lunch, meeting friends or waiting for a tram. Although a considerable variety of play occurs on Bourke Street Mall, not much of this play is generated by its status as a path. Swanston Walk and Southbank Promenade, two of the most popular places to play in all of central Melbourne, are however worthy of closer consideration. These two 1km-long pedestrian axes efficiently connect instrumental destinations for travel, work and leisure. But more than being just mere trajectories, as paths they offer a diversity of experiences to the pedestrian, and they link together spaces and zones of the city which have varied atmospheres and user groups. Swanston Walk runs past the State Library forecourt, Town Hall Plaza and Flinders Street Railway Station, as well as the City Square (refurbished in 2000) and Federation Square (opened in late 2002). There is typically a lot of play on paths like Swanston Walk which offer a multiplicity of opportunities for different kinds of escapism.

Southbank Promenade is another path which opens up the scope of walkers' perceptions and actions by linking together opportunities for play. This path begins at the tight, straight, dimly lit pedestrian tunnel under the Flinders Street Railway Station platforms. It then emerges suddenly onto the wide, open, sunlit, breezy volume of the riverbank and leads onto a pedestrian bridge. From this point, a succession of shifts in elevation and axis intensify sensory and bodily engagement with the landscape (Figure 4.3).

This bridge frames a range of atypical and unconstrained pedestrian experiences which stimulate play. It excites the body's feeling for three-dimensional space. The bridge's platform rises in a continuous arc which lends itself to smooth acceleration on the descent. Its surface of narrow transverse boards provides a pleasant vibration and subtle rasping hum under the wheels of cycles and skates. The bridge's path takes the walker out of ordinary space. It leaps free of the edge of the land and rises high above its surroundings, hanging suspended over the flowing waters where people can feel cool breezes blowing across. The open handrails and minimal structure expose people to a greatly widened perceptual field. The high structural arch of the bridge (Figure 4.4), a very uncommon form in this city, helps to amplify the sense of energetic escape. This bridge stimulates perceptions of vertigo in the most commonly understood sense of the term.

Tire marks from in-line skates and graffiti far up the curve of the arch reveal that people's path from one side of the bridge to the other sometimes leads over the top of the structure, a height 20m above the water. This is a particularly extreme pursuit of vertigo. This alternate path intensifies the height, slope, and the tension between openness and confinement which frames the pleasure of the main bridge deck. People perceive and act out a bodily challenge latent in the bridge's form. The risks are high but the limit indicated by the bridge's handrails is not absolute. The path up and over the arch is smooth and stable. The walk itself need not be too demanding. But the realization that there is no handrail, no margin for error, inevitably raises the stakes. The arch frames a freedom from all constraint, where one has to rely entirely on balance. The player at this game has to call upon all their perceptual and physical resources to achieve the feat safely. The bridge is frequently repainted to remove the evidence of these transgressions, which

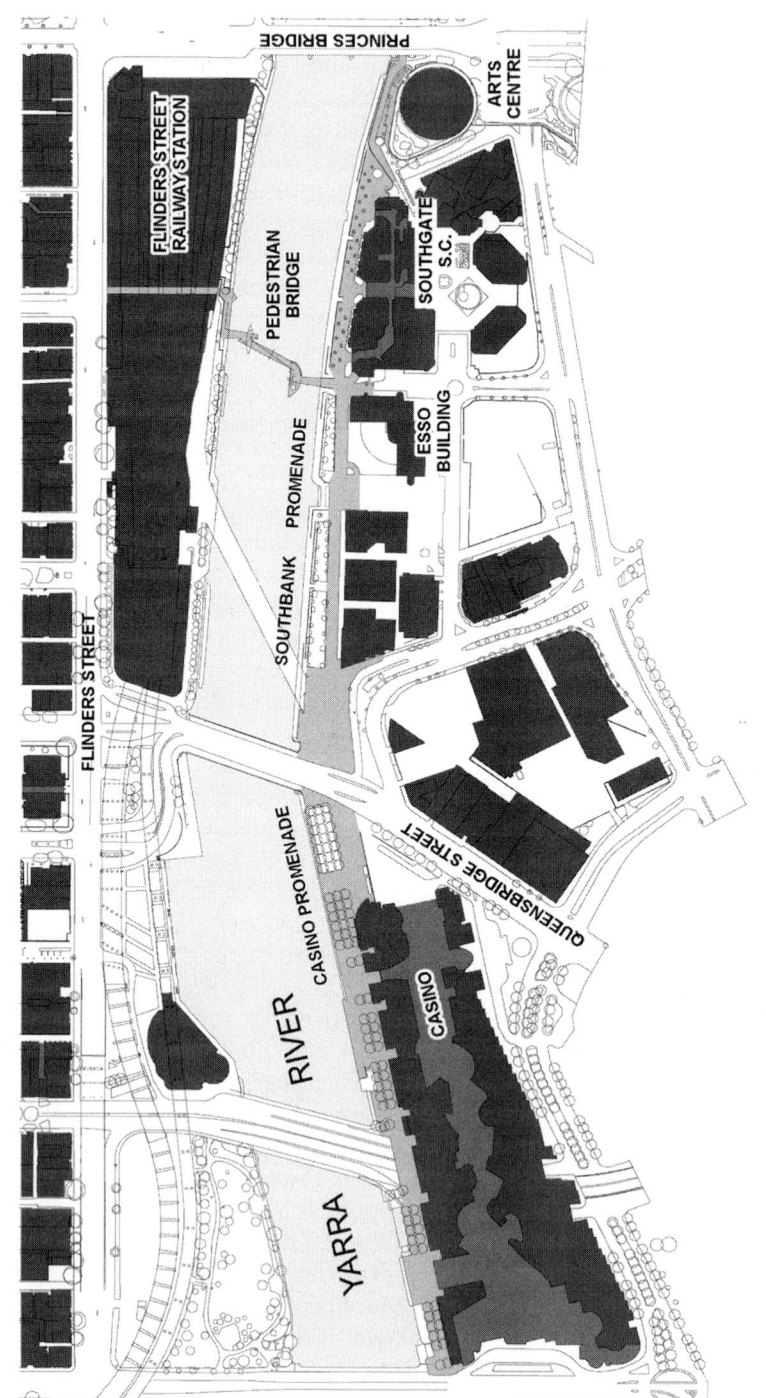

Figure 4.3 Pedestrian path along Casino Promenade, Southbank Promenade and pedestrian bridge, Melbourne.

modify the reality of the space, by marking a new possible path of movement across it (Lefebvre 1991b). For the urban adventurer, such marks show the way through unexplored territory, and also lay down a challenge to those who come after, particularly other graffiti taggers. A vertically grooved surface is later added to make ascending the bridge on wheels more difficult (Figure 4.5), but this too poses a challenge as much as a deterrent, and graffiti keeps reappearing.

On a much smaller scale, daily acts of transgression also often involve pedestrians diverting their path to engage the structural frame of this bridge. People often jump up onto the low triangular stanchion at the end of the arch, or step up on it and slide down the other side. There are many tire marks along both faces of the stanchion, suggesting those traveling on wheels also enjoy this feature.

These uses of the small triangular stanchion and the larger arch show that the attention of people walking along the pedestrian path across the bridge is not completely captured by their perceived goal, the expansive view of the river, their friends or other people around them. Indeed, the presence of others may serve to goad people into displays of brave and skilled engagement with vertigo. People are clearly highly aware of opportunities for playful excess and exploration which lie in their path. This example shows the incorporation of play into the task of walking somewhere. People often integrate playful thrills, tests and displays with their instrumental acts. Playful passage across the bridge is a dialectical development in the playful use of public space, an integration of desire into rational everyday practices.

The bridge links into the middle of the Southbank promenade. The primary attractors at each end of this riverfront leisure precinct, the Arts Centre, casino and Exhibition Centre, focus on spectacular, passively experienced paid entertainment, and the consumptive play of gambling. Pedestrians destined for these venues are seeking escape from rational concerns by pursuing unproductive activities. They are drawn along the long, curving path of Southbank Promenade. The promenade maintains a permanent carnival atmosphere of abundance, energy, fantasy and indulgence. The sheer number of people compressed into the same corridor is part of this thrill. Contributing design attributes include extreme verticality; sensuous shapes (curves, odd angles) and surfaces (smooth, rough); large-scale artwork; brightly colored banners flapping in the breeze; recorded music; jets of flame; leaping fountains and tumbling waterfalls (both of which mysteriously, spontaneously stop and start); and the freedom of extensive, unprogrammed open spaces. A shopping mall can seal in and carefully organize the Dionysian impulses it arouses, so as to maximize purchases. By contrast, the primary attractors, marginality and atmosphere on Southbank inspire play, but they cannot capture or direct people's excess energies. Southbank's diverse distractions and excitations also augment many different kinds of consumption. In a shopping mall, continual distraction by fantastic commodities is a

Figure 4.4 A risky path: pedestrian bridge with rows of metal studs (bottom left) and graffiti, Southbank.

manipulation of people's relatively instrumental needs to purchase specific goods. By contrast, most visitors to Southbank come with the conscious objective of spending a lot of time in a state of distraction.

The character and use of Southbank Promenade vary considerably between its distinct eastern, central and western sections (Figure 4.3). The eastern part of the Southbank promenade leads fun-seeking pedestrians past the cafés,

Figure 4.5 Pedestrian bridge after further modifications to prevent rollerblading, Southbank.

restaurants and specialty shops of SouthGate, which stimulate desires for luxurious consumption. This area is busy with pedestrians at lunchtime and to a lesser degree in the evening, but is relatively quiet at other times. Buskers often perform to the combined audience of seated and passing pedestrians here, although much of this section of the path is too narrow to stage large performances without impeding pedestrian flows; it is conducive to strolling but not to lingering. Much of the remaining width of the riverfront here is taken up and partitioned by tables, planters, kiosks and benches facing in various directions. The mature trees keep much of the width of the path in shade, reducing the visibility of performers. Acts tend to locate in the few pockets where the path widens and the sun highlights the show.

In many respects the central section of the promenade from the pedestrian bridge west to Queensbridge Street offers the best stage. On one particular warm Sunday afternoon, when pedestrian traffic in this area is heavy, a silver woman statue performs on the riverside, while further down a guitarist plays 'Stairway to Heaven' and an artist draws chalk pictures on the pavement. Further to the east, two more women imitate statues: a woodland fairy next to a waterfall, and a copper-colored woman further east beyond the pedestrian bridge. West of the pedestrian bridge, on the steps in front of the Esso Building, an acrobat squeezes himself through tennis racquets.

Everyone moving west along Southbank toward the casino and the Exhibition Centre is funneled through this stretch of path. People who reach this point are in a mood for play, already experiencing an escape from responsibilities and deadlines and hoping to be distracted and entertained. No one is in a great hurry; even the skateboarders generally cruise idly along. This section of pavement is wide and clear of obstacles, and there are few permanent distractions adjacent to it. The river and the city skyline to the north are set back beyond a wide grass verge and largely obscured behind two rows of trees. The buildings on the south side of the promenade are primarily offices, vacant on evenings and weekends. The Esso building is separated by a low wall and moat. On this pathway, people's attention is available, and street performers step up to attract it (Figure 4.6).

Alongside the wide variety of programmed events and semi-professional street performers, a wide range of spontaneous events along Southbank show members of the public as not just consumers but instigators of playful opportunities presented by the physical setting of this path: teenagers doing bicycle tricks; men waving flags in the breeze; children leapfrogging bollards that line the promenade. A young man crosses Queensbridge Street with a crushed drink can stuck on the bottom of one shoe, making a loud scratching, crushing noise as he walks. This too is a form of play: an act done for its own sake, just to see what will happen, a display which draws strangers' attention to him, a heightened sensation of his own motion across the ground. It is an indulgence in the vertigo of disturbing sound, 'rough music' known as *charivari*, performed as a show of derision for social decorum (Darnton 1984). This last example highlights that the act of walking along a path can in itself be a playful experience under the right conditions.

Four men walk shoulder-to-shoulder east along Southbank. Their demeanor is euphoric and their behavior exuberant. As they walk, they spontaneously indulge in many different playful acts, some pure invention and others prompted by people they see. Their walk becomes something of an interactive

Figure 4.6 Adjacent attractions: Southbank Promenade.

freeform procession. They sing a little song. One of the group, wearing a red inflatable ring on his head like a hat, sings the counter-point, with a little dance, waving his arms. When one of the men lights a cigarette, he also lights a tiny firecracker and tosses it ahead of him. There is a sudden very loud *crack* and a small puff of sparks. They come upon a man conducting surveys for the city council, and they enthusiastically crowd in, playfully jostling him. He is smiling, trying to seem in the spirit of things, but his smile seems strained. As the men continue on, they dodge invisible sporting opponents. They hide behind one another as they are walking along and then suddenly leap out. They ogle every woman they pass and chat them up.

Southbank provides a supportive social setting for this adventurous walk. The promenade is a social space apart from seriousness, where few passers-by would take the men's behavior as a real affront. There is a large passing audience of strangers to inspire their play, people who they can catch unawares, and simply pass by if things do not work out. The continual movement of the public through such a space weaves together changing possibilities for social play. People walking in opposite directions along an open, continuous path such as this have a good opportunity to size up strangers. Encounters need only last as long as both parties are comfortable. The absence of car traffic gives people the luxury of moving at their own pace along the path and stopping at whichever displays they find diverting.

The western section of the promenade, in front of the casino, is under the control of the casino management; it is open to pedestrians but not to street performers. The casino management do not use staged performances on the promenade as a way to bring patrons into the area or stimulate a playful atmosphere. The casino itself, in combination with the cinemas, clubs, shops and restaurants which surround it, provide a strong attraction and a stimulus to consumption. The randomized spurting of a fountain built into this promenade, water that constantly cascades down a series of tall black columns, and exploding fireballs which suddenly shoot with a loud *whoosh* from their tops also create an atmosphere of chance, superfluity and excitement on the casino promenade twenty-four hours a day without the need for performers. The anticipation of effortless distraction and thrill is part of what attracts people to the casino promenade.

Nonetheless, a majority of the unprogrammed public play on Southbank takes place along this 500m-long western section. The lack of organized leisure activities here correlates with some very informal, active, risky play behavior. Passers-by can easily be distracted by opportunities for escapism. In this section, the riverside promenade is at it widest. It is relatively easy for cyclists and skaters to move through the crowds, both for the thrill of it and to get to its various playful landscapes, which combine grass areas with generous seating and many sets of concrete steps, ramps and terraces leading down to the water's edge. This is a relatively open terrain, largely unobstructed by trees or sculptures. It is a landscape which encourages playful behavior.

People using this space often have their backs turned to the casino, facing north toward the distractions of the city, the sun and the water. The casino management appear to tolerate active play on its waterfront pathway. A large proportion of the casino's patrons enter the complex from its car parks or by tram along Spencer Street. Many of the people using this stretch of promenade and the play amenities next to it never set foot inside the casino. The majority of those who play in this precinct are teenagers and children. The freedom of play on the promenade can thus be seen as a complimentary attraction to the more organized, adult play indoors. For teenagers, the public condition of the entire waterfront, and the continuity of the riverside path, connecting through to other primary attractors, public transport, the CAD and beyond, facilitate their access to play possibilities in front of the casino.

Changes of use

Most major streets in first-world cities are designed primarily around the needs of vehicles, and they have less of the physical amenities which support and promote play than other public spaces. Yet even in the absence of physical redesign, temporary changes in the function of particular vehicular

streets can enhance prospects for playful behavior. Some of these transformations of use are officially sanctioned. Others are transgressive,
illustrating how playful behavior not only responds to the stimulus of spatial
opportunities, but produces its own space, as a critique of the extreme
functionalism of urban streets.

Official street festivals

In Melbourne, the temporary closures of Russell Street for Chinese New
Year and Lonsdale Street for the Antipodes (Greek) Festival illustrate that
the absence of cars can in itself make a significant contribution to the
possibilities for play. These public festivals introduce many of the play
amenities which have become permanent fixtures on Swanston Walk, the
Bourke Street Mall and Southbank: stages for performance, seating, props
and music. The limited time frame of these events, one weekend each,
contributes to the intensity of the escapist experience. The Antipodes Festival
provides the vertigo of rides and the possibility of competition through
sideshows. The festival also has a lot of information tents: community
organizations as well as companies. People stroll from one end of the block
to the other, taking it all in, exposing themselves to cultural difference. Street
festivals, like busy pedestrian streets in general, gather the city's diversity
together, framing the experience of a 'world' (Lefebvre 1996). Wandering
back and forth between the amusements allows people to experience a
multitude of opportunities and stimuli within a small urban space.

   A much larger-scale reappropriation of more than 40km of streets occurs
in New York City each year during the running of that city's Marathon.
This event has 30,000 participants and 2 million spectators who line the
city streets. Within this specially framed time and space, people find a lot
of freedom not just to run, but to interact with others, and to play. The
costumes of many passing runners in fact provide a dialectical critique of
serious running (Figure 4.7). English police bobbies, with helmets and ties,
mock the orderliness of the event. Superheroes emphasize, in a comical way,
that theirs is a superhuman struggle. Not everyone tries to reach the finish
line as quickly and efficiently as possible: some runners dress in heavy foam
animal costumes. A giraffe costume made at home in Japan and T-shirts
with competitors' own names on them both highlight runners' efforts to
capture attention, to communicate with onlookers when moving past at
speed, and to provoke responses. Sikhs taking part in the race highlight
that costumes are not just for fantasy: they are part of the public display of
identity. The managed spectacle of the marathon helps build tolerance
of social difference. Indeed, it enhances difference, with 12,000 foreign
entrants from 99 countries. This international participation reflects the
historic role of street festivals: gathering people together, an intensification
of the urban condition, and the playful engagements between strangers that
this stimulates.

Figure 4.7 Fun Run: the New York Marathon.

Much play in this event comes through impromptu interactions. Runners are engaged by the crowds, sometimes through shared national identity because both groups carry flags, sometimes by calling out encouragement to strangers whose names are printed on their T-shirts. When runners see or hear friends, they slow down and go over, they stop to say hello. Bystanders do not just consume this atmosphere: they contribute to it.

Higher levels of public involvement can be seen in Berlin's Love Parade, which takes place on a summer weekend along the Strasse den 17. Juni, a key arterial street running through the middle of the Tiergarten, the inner city's major green space. The parade consists of fifty trucks broadcasting techno music from famous dance clubs and carrying dancers in outrageous clubbing attire, many partly naked and some performing sexually provocative acts. The procession makes a continuous linear 3km circuit up and down the street axis, pausing for fifteen minutes at a time. The closure of the street and the circulation of the trucks can be judged as a spectacle, carefully managed and mediated. But this order is not total. People in the audience respond actively to the Love Parade's stimulations, and in doing so participate in creating new opportunities. They make use of the special movement-space which this spectacle establishes to serve their own playful

purposes. The social complexity of the event is evident from different kinds
of circulation which happen within the crowd.

The circling of the trucks brings varying rhythms to the large stationary
audience. Many people in the crowd dance frenetically, and they adjust their
dance steps to suit the beat of the nearest truck. Those who take a liking to
a specific disk jockey follow behind, in groups of as many as two hundred,
although they can easily switch to any other truck as they come abreast
of it and hear the competing sound. The procession is thus fifty parties in
motion, a dynamic flow which picks up, sorts and resorts members of the
crowd, helping those who like a particular beat gather together, and helping
stir the mass. The audience's dancing is stimulated by the spectacle of the
Parade, but is also itself a kind of performance, one which is free-form,
individualistic and meaningful. The Parade is not just a one-way communication of image and sound. The dancing is a way for people to interact,
a form of social dialog.

Another form of active involvement is that many visitors to the Love
Parade come in outrageous and lewd costumes, to make a spectacle of
themselves (Figure 4.8). The visibility which accompanies the greater
spectacle makes their efforts worthwhile. To be seen by the crowd which is

gathered here, these people need to circulate. Many parade themselves along in the large gaps between the trucks. Their circulation involves close bodily proximity, with great potential for engagement with strangers. Generally, people in costume are not afraid to be noticed. Some are masked; their anonymity providing extra freedom for bodily self-expression. Lewd costumes also seem to keep most tourists at a safe distance. The photographing of people's costumes is an active form of tourist involvement. Recording the event changes what the event is: it becomes a fashion shoot, a linear sequence of many brief social encounters. Some photographs of strangers are taken surreptitiously, but in many cases they involve positioning and posturing, conversation, negotiation, entreaties and displays of gratitude.

It has sometimes been suggested that the spectacularization of urban landscapes and civic events sacrifices the local in the interests of the global and the authentic in the interests of the economic, and that this makes places and times more homogeneous (Harvey 1989; Kearns and Philo 1993). But spectacular global events can also stimulate public involvement and enhance particularity, and this is partly because they override the everyday functionality of certain urban spaces, which is itself often a means of discipline.

Figure 4.8 Love Parade: Strasse den 17. Juni, Berlin.

Even though the special street circulation of the Marathon and the Love Parade remains orderly, these spectacles give rise to a wide range of play. Jacobs (1961) and Gehl (1987) both noted that primary, organized activities are necessary attractors to make the great multiplicity of secondary, informal, spontaneous activities in the city feasible. This is true of parades. These events draw people out onto the streets. They provide structure which brings strangers into close contact. They let people see and be seen without being fearful. Although they have a principal speed and direction of motion, they provide a license to greater freedom of action. People shape these opportunities to suit their own needs. Most notably, the parades provide a means for individual identities to be produced and communicated. The Love Parade is a particularly carnivalesque procession which arouses spontaneous, creative and disorderly social behaviors and exchanges which contribute significantly to the form and meaning of the event. The New York Marathon and the Love Parade are special pieces of choreography within Jacobs' (1961) street ballet, which continues unfolding in new variations.

Unofficial street festivals

Sometimes vehicular streets are simply taken over without permission by pedestrians and cyclists to enhance opportunities for playful socialization, as with public events such as protest marches and Reclaim the Streets. In such cases, playful practices become competitions, because they generate tensions with other more instrumental uses of urban pathways.

During rush hour on Friday evening, hundreds of cyclists ride together through the center of Melbourne in a now-global event called Critical Mass. This event is a political critique promoting the right of cyclists to free use of the streets. As the movement's name suggests, the sheer numbers of cyclists in Critical Mass provides a certain power to exercise their claim for space. They appropriate the whole street and ignore traffic signals, suddenly and temporarily blocking vehicular traffic at a series of major intersections. Critical Mass is directly confrontational because it inverts the priority usually given to cars. Tension is maximized by holding the event when drivers are most keen to hurry home and relax for the weekend. By contrast, the cyclists are relaxed and enjoying themselves right there in the city streets, monopolizing the city's most functionalized space. They roll along slowly, chatting to each other, looking around, ringing their bells; they are not really aiming to get anywhere. As a playful escape from efficient transport, Critical Mass is a critique of the belief that both human movement and street spaces should be functional (Lefebvre 1991a, 1991b).

A collective 'bike lift' at the intersection of Collins Street and Swanston Walk (Figure 4.9) highlights that Critical Mass is not serious cycling. This act serves an instrumental purpose. Lifting the bikes in unison is a metaphorical show of their collective strength. The political symbolism of this

Figure 4.9 Show of strength: collective bike lift during 'Critical Mass', Swanston Walk, Melbourne.

act is amplified by its location directly in front of the Town Hall. But both participants and onlookers on the far kerb are smiling broadly, acknowledging their enjoyment of the display. The behavior is playful in a number of respects. As an excessive, frivolous, impractical, fun waste of cyclists' energy, it parodies the practicality of mechanized transport. It is a bodily challenge, stimulated by others who participate and by the expectations of onlookers in cars and on the footpath. The act is exploratory. It is a spontaneous, unexpected behavior which tests a new set of relations between rider, bicycle, space and the public. It transforms bodily engagement: a seated posture, rhythmic leg exercise and smooth horizontal motion in cooperation with a machine is suddenly transformed into a standing clean-and-jerk, the concentrated application of brute force against it. Pulled away in the vertical plane, the bicycle is denied its utility and becomes an encumbrance. This inversion demonstrates an understood reciprocity between bicycle and rider. In terms of the publicness of this display, coming to a standstill also heightens the visibility of the riders, who usually appear to pedestrians as a distant passing blur. The inversion of bicycle and rider also transfers attention away from the contentious vehicle and onto the person who controls it. While playing, the cyclist temporarily becomes a pedestrian. This emphasizes a commonality between the actors and their audience.

The bike lift is also an unexpected and transgressive use of movement space. Vehicles are supposed to stop before an intersection and systematically yield to other's needs, not occupy the space on a whim for their own pleasure. The unproductive, unserious, non-transportational act of the bike lift heightens Critical Mass' general contestation of the rules of the road. The bike lift demonstrates that the accepted spatio-temporal structuring of the street is in fact only provisional. It relies upon an obedient flow of traffic for its constant reinforcement. It can be manipulated by moments which invert or suspend its rules and boundaries. A new form of social-spatial organization is layered onto those which preceded it. The bike lift can be compared to de Certeau's (1993) account of walking in the city: it is a practice which is 'rhetorical'. De Certeau suggests that walking has the same relation to the rules of space that rhetoric has to the rules of language: it manipulates them to create its own logic. The bike lift demonstrates 'the tactics of users who take advantage of opportunities' (de Certeau 1993: 155). Standing still symbolically accentuates the cyclists' intransigence and sense of control.

Critical Mass also has a broader agenda of play, framed by passage through a diversity of urban public spaces and a diversity of actions which arise through these situations. On one particular Friday evening, narrow, pedestrianized Hardware Street is full of café tables which are filled with people at leisure. Suddenly the space is also crammed with cyclists, barely able to keep moving in the congestion (Figure 4.10). Compression within the laneway engenders engagement between patrons sitting stationary on the edge and those passing through, without great effort from either party. It makes spontaneous interaction less demanding. The route of Critical Mass cuts across the complex organization of the city streets, weaving in a new pattern of social behavior. The event follows a new route each month, unknown even to most of the participants. This strategy serves three distinct needs. It allows them to put themselves on display to a different cross-section of the public. Moving along this narrow lane puts them in close contact with non-cyclists, some of whom smile, wave and make supportive comments as they pass. Second, it allows them to experience new parts of the city, including streets and lanes which are not generally open to bicycles. Their growing understanding of the urban street network is essential for freedom in using it. The path through Hardware Street offers many distractions, including the escapism of live music put on by café managers. The cyclists' wandering path also frames various different experiences of vertigo, from the testing thrill of squeezing through a tall, narrow laneway shared with pedestrians, to the exhilaration of rolling swiftly and smoothly down the long, broad slope of Bourke Street. Third, Critical Mass' strategy helps them elude the police, who seek to predict and regulate their journey, to minimize its potential for social friction.

Critical Mass is a rich and flexible combination of tension, critique, expression, exploration and invention. The fluidity, multiplicity and spon-

Figure 4.10 Spontaneous appropriation of a street designed for leisure: Hardware Street.

taneity of its new practices emancipates participating cyclists from the functionality which is mapped onto urban streets (Jackson 1988). Altered use of the street network sets a model for how the city's existing system of paths can be used differently, to frame new social and bodily experiences of play, and to perhaps even reshape the definition of appropriate behavior in such spaces.

Changes of mode

Critical Mass is an example of how change in the use of a street or laneway arises out of a change in the way of traveling along it. There are many other forms of playful travel through the city which enrich the diversity of people's experiences of urban paths.

Early one Saturday afternoon, a family with two young children are touring through the city, all wearing in-line skates. The father pushes a third child along in a stroller which has large spoked wheels. They pass through Block Arcade, a nineteenth-century retail development, cutting in and out of the pedestrians (Figure 4.11). Ahead of them, a group of older people standing around the edges of the Arcade's rotunda applaud a flautist and

Figure 4.11 Not promenading, gliding: Block Arcade, Collins Street.

harp player who perform a sedate concerto. This scene acutely frames a contrast between passive and active attitudes toward play. Rather than leisure through receiving pleasure, the family produces play. Rolling is vertigo, it is a different way of moving through and experiencing the city. This family, like the pedestrians in the arcade, are taking advantage of the quietness, the safety from vehicular traffic, from weather and from crowds on the streets. They are also taking full advantage of the smoothness of the arcade floor, which makes possible the sensory experience of going past the arcade's fine boutiques at speed. The family discovers and enjoys new experiential opportunities that lie latent in this environment. Their playful act reveals and performs an extra potential of this urban pathway.

Two adult in-line skaters race each other in circuits of the narrow plaza around the base of the Melbourne Central office tower just as people are leaving work on a Friday afternoon. They move fast across the smooth paving, jumping off flights of steps, swerving between trees and pedestrians. The risk and the thrill of moving fast is here enhanced by the obstacles the urban space presents, including the tightness of the space, changes in level, the many workers exiting the building, and by the limited reaction time when coming around the building's corners.

Such playful modes of transport offer a critique of instrumental, efficient, easy movement through the city toward some fixed destination. For these skaters, getting there is most of the fun. This is travel as a consumption experience, a form of tourism which is focused more on the feel of the ground than the spectacular view. Most tourists perceive a city quite passively, but for people on skates the urban landscape is not consumed as pre-packaged (Percy 1975). Their leisure behavior escapes the codification of leisure in the practices of promenading and sightseeing (Rojek 1995). Skaters reread a path which the average pedestrian takes for granted. Traveling on wheels reshapes the skater's perception of the city: it is more active, intense, immediate, risky, subtle and ephemeral than that of the walking sightseer.

Skaters' experience of the city is compressed in time and space. Skaters' velocity demands constant focused attention and precise coordination; the senses and body are stimulated by rapid, intense engagement and sudden encounters. Skaters have a heightened awareness of the city's surface geometry. They focus on the swift and pre-emptory reading of texture, inclination, and elevation ahead. They explore the potentials which paths through the city provide for acceleration and effortless coasting, for friction, for leaping free of the ground and sliding: in short, for the thrill of vertigo.

Skating play occurs in several distinct areas of the Melbourne CAD. There is a cluster of popular sites around the wide non-vehicular paths of Swanston Walk and Southbank. Another two popular areas are less immediately obvious: the south end of William Street and the east end of Little Bourke Street. These latter areas have two clear benefits for skateboarders. They include a number of smaller plazas and laneways set apart from vehicular traffic, where skaters can control the kinds of risks they are exposed to. William Street is the heart of the city's financial district, and this area is primarily used by skateboarders on evenings and weekends when its many paved surfaces have been abandoned by office workers and security guards. The other key feature of these areas is not obvious from a street map – which is after all an abstract, practical device – but becomes body knowledge to those who traverse cities on wheels. The reason skaters play in these locations becomes apparent when the locations are overlaid on a topographic map of the city center. These are two areas within the CAD with sloping topography, where much of the land has a gradient of over 7 percent. Such paths as William Street and Little Bourke Street frame opportunities for downhill acceleration, which brings the thrill of vertigo and also facilitates movement from one plaza to the next. The same is true for the north end of Swanston Walk. In Figure 4.12, skateboarders progressively move down the block of Swanston Walk in front of the State Library, testing their skill against ledges, sculptures and benches, trying out different tricks. The continuous slight downslope adds to their experience of Swanston Walk as a sequence of opportunities and challenges which draws them constantly onward in search of new thrills. Slope is just one example of the many sensory characteristics of urban paths which most people register only subconsciously.

Figure 4.12 Skaters move along the downward slope of Swanston Walk.

Skaters' trajectories are new ways of weaving the diversity of the city together. Their turning, jumping and sliding creates alternative routes, their play stitches together new contours and new sequences of meanings (Borden 2000). By tightly focusing most attention on movement as an end in itself, skating provides an escape from rationality. Skating is not always efficient, but it is often enjoyable:

> as well as being transport and recreation, skating is also travel. Here we mean travel in a sense qualitatively different from transport. Travel is not the maximizing and efficient act of moving from A to B, but rather is the satisfying pleasure of experiencing place, of celebrating the means and not the ends.
>
> (Stratford 2000: 10)

As the endless repetition of moves by skateboarders suggests, they do not always move through the city in order to get somewhere. Their path is often uneven, circuitous and risky. But moving can in itself be a sensory delight and a bodily challenge; an escape from normal ways of viewing, traveling through and inhabiting the city. The smooth, hard, continuous surface of urban footpaths and plazas is ideal for journeys on wheels and offers a multitude of possible paths for play. Playful ways of moving along urban pathways often involve risk, stretching the limits of experience, testing one's skill. Each act is a new, relatively unfamiliar encounter with an urban landscape and with the diversity of people who move along it. Movements on wheels compel people to reread urban space, producing the terrain as a new kind of experience and drawing out its potentials. These examples offer general insight into opportunities which lie open to all those who move along an unfamiliar route in the city, even if only walking.

Changes of meaning

Critical Mass illustrates conflict between serious and playful practices when their paths cross, and when paths used for play preempt other more serious demands on their use. This is competition: the play action of such processions is a direct challenge to instrumental function. But tensions can also arise between playful and pragmatic processions in the mode of simulation. In such cases, playful behavior appropriates, critiques and expands the social meanings which are written into urban space by ritual procession along paths.

Swanston Walk is the path of most of Melbourne's public processions. Major institutions located on this axis including the State Library, Town Hall, St. Paul's Cathedral and the Arts Centre, and it is terminated by the city's main war memorial. Ritual movement along the axis draws these various sites together into narratives which reproduce and reinforce cultural beliefs, and bind social identity to place. This street is the scene of constantly evolving tensions between formal civic processions, popular marches and carnivals and instrumental needs of commerce and transport, as well as a wide range of transgressive practices. The city's most popular annual procession along Swanston Street is the focus of a festival, Moomba, created by civic leaders in the 1950s. Historically this took the form of a parade of decorated floats. The Moomba parade had supplanted the older Labor Day parade, which celebrated the proclamation of the eight-hour working day. Moomba had nonetheless eventually come to be seen as the 'people's festival'. Swanston Walk, the Moomba holiday and the parade are at the center of a dialectical struggle, with authorities attempting to channel the surplus time and energy of the urban population away from excessive behavior and disruptive, destructive or transformative possibilities (Brown-May 1998). The pedestrianization of Swanston Street and the redirection of vehicular traffic in the 1990s can be seen as an attempt to manage the impact which frequent celebratory processions along this axis, including Moomba, had on instrumental functions of transportation in the CAD.

In the year 2000, the Moomba festival parade is replaced by a procession of decorated trams containing professional performers. The themes of the trams are playful reinterpretations of aspects of local urban culture, and evocations of the abandonment and social transgressions and inversions of carnival. Traditionally, the Moomba parade had centered on active participation by community groups: hundreds of costumed people marched and danced along, accompanying thematic floats they had decorated themselves. The parade had displayed and invigorated the reality of the city's ethnic and social diversity; a marked contrast to 2000's symbolization of it. This year thus marked a tightening of the regulation of public leisure time. A participatory public celebration was replaced by a spectacular simulation which people were supposed to passively watch. This spectacular depiction of diversity and fun serves to mask the production of behavioral controls. The medium – in this case, the social space of the parade – is part of the message: the public become marginalized in the role of passive spectators. Public leisure is carefully choreographed, on the very day intended to sanctify the idea of public 'free time': the Labor Day holiday. This transformation is a clear example of public life in contemporary spectacular society, where 'what was once intensely lived becomes mere representation' (Debord 1994: 12).

However, even in this context, the spatial actions of certain individuals succeeded in freely expressing different social meanings. In playful counterpoint to this parade, a small protest march moves down the footpath of Swanston Walk at the same time as the decorated trams moved along the middle (Figure 4.13). This march features a person dressed as a giant budgerigar. It looks like part of Moomba, but the protesters accompanying

Figure 4.13 Contesting the spectacle with the body: Birdman rally next to street full of parading trams, Swanston Walk.

it chant 'Bring Back the Bird!' They seek to draw attention to the cancellation of another traditional part of the Moomba festivities: the Birdman Rally. In this competition, members of the public launched themselves in home-made, unpowered aircraft off the side of a city bridge and attempted to pilot them over a set horizontal distance. Many participants took off in nothing more than a funny costume, wildly flapping their arms in a ludicrous imitation of flying. The ungainly bird maquette ably represents the whimsical spirit of this contest. The Birdman Rally was a grand example of public play. It was a participatory event which brought together competitive display, intense, risky experience of the body in space, and sudden, dramatic wasting of energy. Although broadly regulated, it promoted freedom, the pushing of the limits of human experience. The budgerigar, a common household pet, is an ideal rallying symbol of this quest for free flight.

The protesters struggle against the curtailing of behavioral excess at Moomba. They do so by harnessing the social setting of the formal proces-sion and turning it against itself. They appropriate the audience gathered along the path of the main parade by running their own event parallel to it. Their march simultaneously challenges the main parade and attempts to look like a part of it, through the use of a giant, fun, colorful figure. Their performance also offers a critique of the organized parade as a passive consumption experience. Moving along the footpath, the group are closely engaged with the crowd. They invite people to join them. Their display is unanticipated and their intention is somewhat obscure. They bodily reinvent the parade as an active celebration of nonsense, imitation and vertigo.

A second case of simulation which involved parallel movement along a path occurred on Australia Day 2000, the public holiday celebrating the founding of the nation. Outside the Town Hall a dense crowd gathers and important people make speeches; confetti and streamers cascade down in an explosion of national fervor. Hundreds of tiny flags are waved patriotically in time to a traditional tune about the first white settlers of the country (Figure 4.14). Then an official 'Millennium March' heads south along Swanston Walk from the Town Hall across the river to the Alexandra Gardens. The 1000 walkers are kept in a tidy group. Five minutes later, uniformed staff are disassembling barriers and sweeping the street clean, removing all trace of the celebration. The event is allowed only to minimally disrupt Swanston Walk's main 'function' as a street.

An hour later, an Invasion Day march, protesting against the nullification of the rights of Aborigines and their mistreatment at the hands of European settlers, departs from the central Post Office in the Bourke Street Mall and turns along Swanston Walk, retraces the route of the Australia Day march, passes by the Town Hall without any sign of recognition, and ends at an Aboriginal sculpture installation in the same gardens (Figure 4.15). The Invasion Day march draws symbolic power not by confronting the Australia Day march, but by engaging with the path where it took place. It displaces

Figure 4.14 Street as national identity: Australia Day celebration, Swanston Walk.

Figure 4.15 Rewriting a path: Invasion Day march, Swanston Walk.

the memory of the earlier march along Swanston Walk much more effectively than the cleaning crew ever could, because it inscribes Swanston Walk
with a new meaning, highlighting that this is also, and has always been, an
Aboriginal path. By retracing this same route, the second, simulative march
demonstrates that this symbolic terrain is both physically and ideologically
contested.

While the Birdman Rally march is adjacent in space, the Invasion Day
march is adjacent in time. In both cases, parallel play on Swanston Walk is
a means to playful simulation of ritual movement along that path. The two
transgressive parades use simulation to dissimulate the authority which has
been marked onto this axis by the formal processions, harnessing their
symbolic power, utilizing their conventions of movement and meaning and
their concentrations of people and emotions. The playful parades steer this
path to new purposes.

In some respects, the official processions seem to be playful. The Moomba
and Australia Day marches provide a framework for non-instrumental
interactions between strangers. They are stimulating to the senses. But these
official processions carefully regulate the roles and locations available to the
general public; little is left to chance. They lack an element of freedom. The
comparison highlights that public processions provide very controlled,
ritualistic forms of escapism, because their chief aim is to reproduce a certain
image of community in the space of Swanston Walk.

The secondary marches themselves also have instrumental goals: to raise
social awareness and encourage resistance. But they can employ a greater
flexibility in means. Their very aim is to undermine the decorum and sense
of social order which repeated formal processions help to reproduce. While
they are no less organized, these secondary marches engender more spontaneity and creativity in their execution. These playful parades and their
participants are open to a greater range of involvements with strangers. They
have an air of casual enjoyment, inclusivity and tolerance which the official
processions lack.

The route of a parade constructs a narrative which links understandings
of place and self. But if social identity can be spatialized through action, it
can also be constantly reimagined and re-written in time and space. The
Birdman Rally and Invasion Day marches illustrate playful construction of
new 'processional texts' on Swanston Walk, even though they pass by its
symbols in a familiar sequence (Brown-May 1998). Indeed, their playfulness
thrives on a dialectical negotiation with predominant forms of representation
which they seek to exploit and conquer.

The shifting flows of play

The creation of large-scale, comfortable, traffic-free promenades, walks
and malls is one way to organize pedestrians' experiences of the city into

perceptually and socially stimulating sequences in time and in space, and to thereby encourage a wide range of leisure activities. Short-term street festivals, both official and informal, suggest that temporary closures of vehicular streets can be highly stimulating for play. It is a general characteristic of perception that a person's reaction to a stimulus decreases as a stimulus keeps repeating; people are naturally more aware of new sensations. Street festivals suggest the general urban design principle that novel, temporary modifications to urban spaces can help to stimulate playful exploration of opportunities; to serve play, urban spaces should be transformed more frequently than instrumental function necessitates. Critical Mass provides interest and challenge for participants and gains public attention both because it interrupts the normal functional flows of the city and because it keeps exploring new paths.

Because play is stimulated by diversity, playful behavior thrives in areas where there are choices of different types of path – vehicular/pedestrian, indoor/exposed, wide/narrow, direct/meandering – and also choices of route. Not all play is about comfort and acceptance of planned conditions. A variety of playful experiences of vertigo are inspired by more engaging and challenging opportunities for action which particular paths provide: the constriction of laneways; the height and openness of the Southbank pedestrian bridge; smooth surfaces and slopes which enhance speed; or, alternatively, rough surfaces such as the stanchion of the bridge which is treated as a hurdle and the drink can jammed under the teenager's foot.

The diversity of oncoming people encountered along one's path also serves as a stimulus to play. The New York Marathon and the Berlin Love Parade create special patterns of flow which give rise to a wide range of play among strangers. The funneling of pedestrians at leisure onto the single route of Southbank Promenade frames a large audience which encourages numerous forms of display. Conversely, the many narrow, winding routes through laneways between Flinders Street Railway Station and Bourke Street Mall create the potential for many different chance encounters between individuals. Prospects for choice, novelty and chance encounter are particularly important for encouraging play by the everyday commuter, whose path must always start and end at the same point.

People can also escape the functional, efficient ordering of human movement through urban space by changing their mode of travel. This can compress and accelerate or prolong experience of the city and of other people. Skating and cycling illustrate that moving through the city need not be instrumental, but can be a richly stimulating experience. The design characteristics of paths provide possibilities for these new ways of moving along them. Skating is a particularly tactile form of travel which works both with and against the specific material conditions of the terrain. Skating is in sharp contrast to conventional tourism, with its passive, distanced consumption of landscape as image.

People's movement along a path is also influenced by their awareness of the meanings and the history of the setting. Repeated, ritualized use of an urban pathway makes it into a repository of meanings, which it lends back to the social practices that occur there. The Birdman Rally and Invasion Day marches provide playful, transgressive commentaries on the existing narratives that are embedded in the paths they trace. These playful parades show that people engage with spatial meanings that suit their fancy; they ignore some representational contexts and they turn against others. Meanings certainly adhere to the built environment, but they cannot be fixed there and cannot easily be enforced. These acts of simulative play reveal the inalienable freedom people have to take advantage of opportunities to explore and redefine the meaning of places.

People's playful actions constantly develop the uses, modes of encounter and meanings of urban pathways. The creativity of street experience confirms de Certeau's (1993) depiction of the city as a continually unfolding story which is written by the diverse trajectories people take through space. There are as many paths as there are acts of walking. A diversity of playful movements along paths constantly rediscover and redefine the substance of urban space as a social artifact: 'Their swarming mass is an innumerable collection of singularities. Their intertwined paths give their shape to spaces. They weave places together . . . they spatialize' (de Certeau 1993: 157).

Chapter 5

Intersections

Intersections exemplify Lefebvre's definition of urbanism:

> The form of social space is encounter, assembly, simultaneity ... [it] implies actual or potential assembly at a single point ... Urban space gathers crowds, products in the markets, acts and symbols. It concentrates all these, it accumulates them. To say urban space is to say centre and centrality.
>
> (Lefebvre 1991b: 101)

At intersections people are exposed to the greatest density of other people and the greatest range of sensory phenomena and opportunities for action. Where paths intersect, people are brought up close. It is common to encounter strangers who have different trajectories. Because of restricted visibility, these encounters can happen quite suddenly and unexpectedly. Hence intersections can be experienced as a compression of social time and space. This intensification can stimulate playful responses. Intersections also punctuate journeys through urban space. People's need to change direction, or to navigate their way through intersecting flows of people or traffic, generally causes them to slow down, or to become temporarily stationary. This intensifies their attention to things around them, the things which are not in their line of movement (Lynch 1960). It also increases the opportunity for engaging with these distractions. An intersection expands time, creating a time apart during which play is possible. Intersections are points of both convergence and divergence. Intersections broaden the field of vision, opening up new options for experience and directions for movement. Phenomenologically, space opens out at an intersection. Intersections are thus sites where people can be distracted from instrumental purpose.

The intensity of movement and the level of distraction at any particular intersection are products of both its physical structure and its relative centrality within the wider pedestrian network. Differences among intersections can be illustrated through a series of examples from two dissimilar sections of central Melbourne: Swanston Walk and Southbank.

Swanston Walk runs approximately 1km north–south through the heart of the prime retail area in the Central Activities District, a nineteenth-century colonial grid of 200m × 200m blocks which are all split by east–west-running minor streets which have become popular retail areas (Figure 4.1). The character of Swanston Walk thus differs from east–west-running major streets in the CAD because of its shorter 100m blocks, similar to New York's north–south avenues. Major train stations are located at Swanston Walk's northernmost and southernmost intersections within the city grid. Several major public institutions are spaced along the east side of Swanston Walk; retail uses line the west side and stretch for at least a block both east and west.

Southbank Promenade, including the casino promenade, runs for approximately the same length (1km) as Swanston Walk, but in an east–west direction. Starting from the west end, road bridges across to Southbank break it into blocks of 150m, 250m and 600m. The structure of blocks and intersections is very complicated. The first intersecting street from the west end, Kings Street, forms an overpass at the riverbank. The casino complex was built both over and under this overpass, and presents a continuous frontage of more than 500m to the river. The second bridge, Queensbridge, bends and divides to link to two wide traffic arteries, forming an awkward, exposed junction area approximately 200m across. One of these links has now been closed and a new mixed-use cluster of large buildings is nearing completion, but observations which follow in this chapter describe the earlier configuration. Further east, a pedestrian bridge splits the last long block into two roughly equal segments. At the eastern end, Swanston Walk itself bridges the river 10m above Southbank Promenade. The bridges have allowed the redeveloped Southbank to 'plug in' to Melbourne's pedestrian-friendly CAD by roughly replicating its standard block pattern. But this is not a replication which continues any further south; these links are intended primarily to bring people only as far as Southbank. It is technically possible for pedestrians to continue further south, but there are fewer links, and the legibility and the general quality of the pedestrian environment are very poor, with few active frontages and many set-back buildings and parking garages. There are far fewer pedestrian-oriented activity areas beyond Southbank, and this has a significant impact on the kinds of activities that are likely to occur on the intersections along Southbank Promenade.

Encounter

At urban intersections, people with different needs and interests, moving in different directions, are concentrated together and must share a tight space. This frames the possibility of encounters between strangers. An instructive case is at the northwest corner of Collins Street and Swanston Walk, where a man often stands holding a large signboard (Figure 5.1). His positioning indicates that he is keen to engage people. Attracting the attention of passers-

Figure 5.1 Intersection as visibility: Swanston Walk at Collins Street.

by depends on visibility, both from a distance and up close. People have heightened perception of objects and opportunities at intersections because they have to make decisions there (Lynch 1960). The man frames himself so as to capture that attention. At close quarters, he blocks direct passage, but even from a distance, the man's sign interposes between the pedestrians and their view of the path ahead. When he stands exposed just beyond the edge of the building's awning, his sign is quite luminous against its dark, enclosed understorey. The expansiveness and relative brightness of the sky-lit volume of the intersection serves to highlight whatever occurs there. The contrast is sharpest at this particular corner, where the waiting pedestrians often gaze idly out over the open space of the City Square diagonally opposite. The challenging, enigmatic messages on the sign are framed between the immediate bodily discomfort of the crowd and the possibility of staring blankly into space. By standing out in the middle of the footpath, the man maximizes the sign's visibility to people approaching from the other corners of the intersection and further along the streets. In the Melbourne CAD, very few buildings are set back; facades are generally aligned with the front property boundary. Such an intersection is one of the few points where such a display is widely visible from more than 30m away.

This man does not stand to one side, but directly in the flow of traffic, on one of the city's busiest intersections, where 12,000 pedestrians pass during the lunch hour (City of Melbourne and Gehl Architects 2004). Whyte (1988) observed similar outcomes in New York City. The passing crowds are an essential element of this opportunity to express his opinion publicly. It is at such intersections that he can make contact with the greatest diversity of people, to generate friction, to stimulate debate. Every once in a while, someone stops and steps forward to engage him (Figure 5.2). There is no way of knowing who they might be or guessing their point of view. This element of chance, an escape into an unpredictable, relatively unrestricted social involvement, is part of the thrill of standing out there. Engagement with this unusual individual is also fun for passers-by. He distracts them from their everyday concerns and predictable social relations. His sign is an invitation to open an encounter (Goffman 1972, 1980). Passers-by have a certain amount of control over when and how to engage him. They are free to leave when the discussion no longer gives them pleasure.

Yet the density of pedestrian traffic, and the way the footpaths meet at right angles, at a built-up, 'blind' corner, means some engagements with the man and his sign are close and sudden. These involvements are unplanned, and therefore generally without instrumental purpose. They arise spontaneously from sudden conjunctions of people and the circumstances of their orientation and motion. Strangers generally keep their distance in public through a carefully managed combination of attention and 'inattention' to those around them (Goffman 1971, 1980). However at an intersection, a person is often brought into close proximity with both stationary people and people who are moving through between them on the cross-street. Many people cross one's line of movement and pass across one's field of vision. Their proximity leads to intense visual, auditory, olfactory and even kinesthetic perception of them (Rodaway 1994). The compression of many people into the small space of a street corner brings about the transgression of personal boundaries. There is a necessary intrusion of many strangers within the social distance of 2.1m, such that individual verbal contact is possible. At urban intersections, it is to be expected that the bodies of strangers will pass within the personal distance of 0.8m, sometimes even closer than the intimate boundary of 0.5m nose-to-nose separation (Hall 1966). The stress of this proximity, in terms of the demands it places on careful face management, is mitigated somewhat by the fact that people are usually either standing side-by-side or approaching each other within their peripheral vision, rather than confronting face-to-face. Because of this orientation, the gaze and body of other people suddenly have to be negotiated at very close range.

The traffic signals at this vehicular intersection provides the man with the signboard the advantage not only of the compression of space, but also of a concomitant distension of time. While pedestrians stand waiting for the

Figure 5.2 Intersection as encounter: Swanston Walk at Collins Street.

signal to cross, they are forced to pause the practical task of hurrying along the street. They become a captive audience. They have a few free seconds during which their attention is open to distraction. Many glance up and read the man's statement. It is hard to know what to make of it; it challenges their frame of mind. The sign man's intervention creates its own time economy. In a space where people are initially forced to halt by the pragmatic technology of traffic lights, they sometimes choose to linger, even if it is mid-afternoon on a work day, and to indulge in discussion in the midst of a flow of strangers. Most of the people who engage him stay and talk even after the crossing signals have run a full cycle. While he is busy talking to a suited businessman, another young man with dreadlocks is willing to wait his turn to ask questions.

Waiting

Whyte (1988) observed that the timing of traffic signals clusters pedestrians together at intersections, because they must wait there. With a traffic signal cycle of 100 seconds and 140 persons per minute walking along Melbourne's Swanston Walk at peak times, divided across two-way flows on each footpath, approximately 25 people become clustered together waiting for every single change of the signals. This is a time apart, when people have to stop the practical task of walking somewhere. They are freely available for involvement in things that seem pleasurable. During the time the signal is red, people have time on their hands, time when anything could happen, anything could catch their eye, and other people are within view. In the same 50-second waiting period, an average of 25 people will walk past in both directions along the same side of Collins Street (City of Melbourne and Gehl Architects 2004). The pause at the signal provides opportunities for unanticipated engagements. There are tram stops in the CAD at each intersection along nine of the major streets, and waiting passengers are also potential audience or participants for any playful action.

A person wearing a cow costume and carrying a pile of advertising flyers is waiting in a large, tight crowd to cross Flinders Street into the train station. A stranger who has just crossed Swanston Walk with a friend also wants to cross Flinders Street. He walks around behind the cow and surreptitiously pulls her tail, then quickly steps around to the opposite side and stands abreast of the cow, blank-faced, as though he has been innocently waiting to cross the street. The cow spins around the wrong way to see who has pulled her tail – but not quickly enough. Three boys sit together on a tram stopped at Melbourne's busiest pedestrian intersection, Swanston Walk at Bourke Street Mall. One boy starts to wave wildly and smile out the window, hoping to catch an eye among 60 people standing at the adjacent tram stop. His action is so caricatured, he is obviously mocking unreserved friendliness. His separation within the vehicle, which will soon move swiftly away, limits the risk of this playful encounter. At another intersection a couple talk while they wait for signals to change. He is apparently teasing her, because although she is speaking cheerily, she often punctuates her comments by giving him a loud and exaggerated slap on the shoulder. When the signal changes they cross, holding hands. A woman waiting for the signal to change at a different intersection grabs her partner by the head and plants a kiss on him.

Such examples show people engaging with others – either strangers or companions – while they wait at intersections. These are not intense confrontations; what unites these examples is that they take advantage of a time which must necessarily be spent at that location, a time to be killed, consumed without serving a productive goal. Playful activity fills the temporal void at the intersection, a void which has been created to suit the instrumental demands of traffic. Waiting at an intersection space does not always mean standing still. Each of these instances shows some basic characteristics of play: action which is spontaneous, exuberant and non-instrumental. Passivity brought to bear on the act of walking is transformed into an explosive release of energy, a freely constituted engagement with other people, framed by circumstance. The man pulling the cow's tail highlights the flexibility of the pedestrian. He can easily change direction and velocity and combine other activities with his walking. Walkers are able to move faster or slower in response to interests and tensions which the conjunction of paths might present, to shape an unfolding encounter into a game. For this play to be effective requires that the intersection has enough space for the man to make his maneuver. The blinkered vision of the cow costume is emblematic of the single-mindedness of many pedestrians and the tightly framed directionality of the urban footpath which serves their pragmatic needs. This goal-focused demeanor heightens the tension of any peripheral interjections from strangers.

The number of junctions in the Melbourne city center where pedestrians have to wait for vehicles moving across their path is particularly large, because of the multitude of mid-block arcades and minor laneways within the street grid, and the traffic closures of Bourke and Swanston Streets (Figure 4.2). Alexander et al. advocate these kinds of interlaced systems, with 'two overlapping orthogonal networks, one for roads, one for paths, each connected and continuous, crossing at frequent intervals' because 'a great deal of urban life occurs at just the point where these two systems meet' (Alexander et al. 1977: 271–72). Play appears to be one of the aspects of urban life nurtured by such a network, because playful encounters become possible when pedestrians have to wait.

Distraction

At intersections, the horizon of a path suddenly open out, offering new directions. The crossing of paths brings exposure to new and unfamiliar sensations, and also frames unexpected encounters. The diversity of stimuli gathered at an intersection can draw people's attention off axis and away from the efficient carrying out of predetermined instrumental objectives. Intersections also frame a stationary audience, when traffic lights cause the suspension of practical action. These conditions support simulative play. Performances and representations rely on attracting the roving eye in a liminal instant when it is not consumed with the function of safe passage.

Urban intersections have always been popular sites for staged displays, including street performers, public speakers and advertising billboards. But the anonymity, complexity and freedom of behavior in public space means that the effect of such stimuli cannot easily be predicted or controlled: play often happens by chance. Caillois (1961: 73) suggests that 'no simulation can deceive destiny', arguing that simulation and chance are incompatible forms of play. But the circumstances of social engagements at intersections only begin from chance, and unfold in time as simulation or competition, as attempts to enjoy the unexpected. The following examples reveals a range of ways in which people, once distracted, take control of randomly encountered events, and attempt to 'make use' of them, in a playful, non-instrumental mode.

A man creates art on the footpath, drawing a cigarette in chalk, then scattering a large pile of butts in the shape of the smoke (Figure 5.3). He adds the caption: 'Do your lungs look better?' People stand and watch as he creates it. Only some passing pedestrians notice the finished work; several people step on it before they become aware of it. In the swirl of pedestrians, the dipping and turning of one head prompts others to do the same. One man waiting for the lights to change looks back over his shoulder to see what is going on, examines the artwork, and an incredulous look comes across his face. Another pedestrian next to him then also looks back, and the first man makes some comment to this stranger. A young man points the artwork out to his companion, then mocks an emphysematic cough. Although engaged in getting from one place to another, he becomes actively involved in simulative play.

This example shows how an intersection frames possibilities of concealment as well as visibility. Because the footpath is crowded with people both stationary and moving in other directions, and because the artwork lies at one's feet, below the normal field of vision of the urban pedestrian, those walking past encounter it quite suddenly. Hence their responses are spontaneous.

Another case of distraction occurs on Southbank's riverside promenade at its pedestrian crossing with Queensbridge Street. Three tourists walking along the promenade suddenly come upon the view of an enormous advertising sign in the form of a brewery's logo, which frames the south side of this wide intersection (Figure 5.4). They are so excited about this sudden discovery, one man takes a photograph of the other two, posed drinking

Figure 5.3 Intersection as distraction: Swanston Walk at Collins Street.

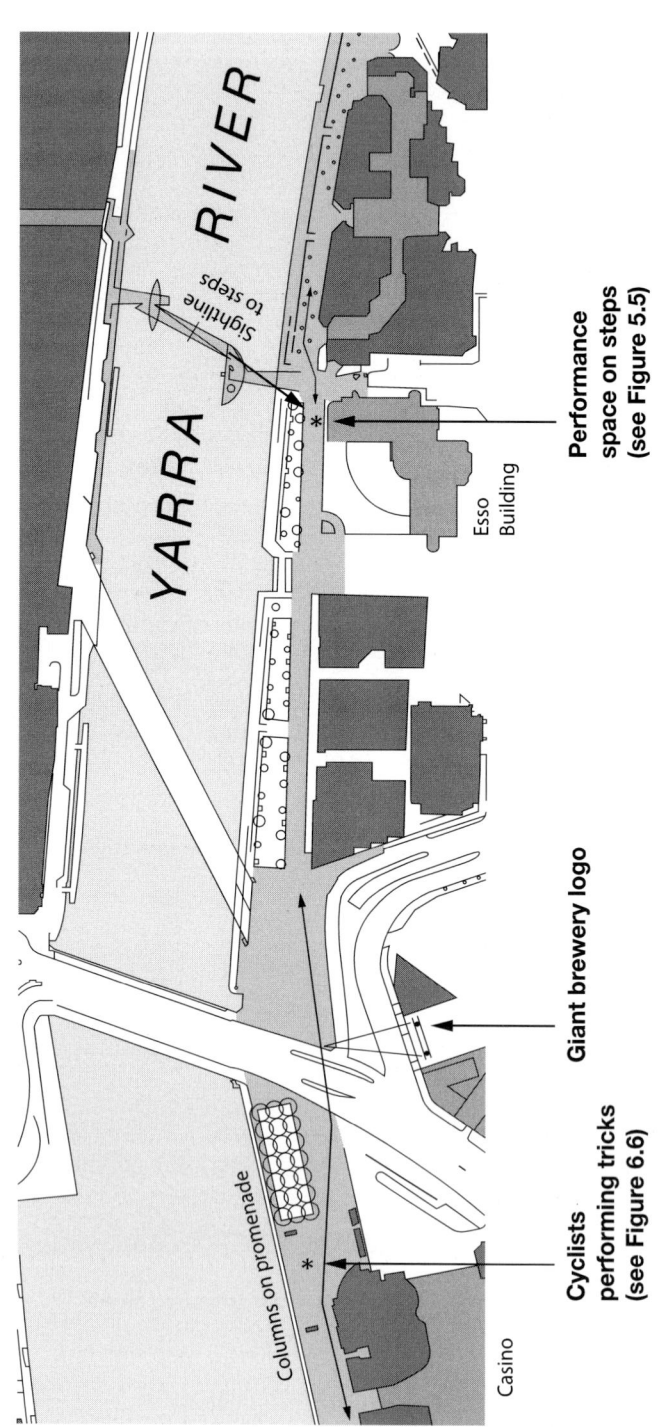

Figure 5.4 Different playful potentials at various intersections along Southbank Promenade.

from bottles of the brewery's beer in front of the giant logo. This intersection has framed an unexpected encounter with a familiar image which has powerful resonance in their lives, but which takes an unfamiliar form. The sign is so large it is overwhelming; it stimulates a sense of vertigo. The perspective of a photograph from this point frames the men and the logo at the same size. This photograph thus captures an attitude of playful reciprocity (Gilloch 1996). Through the matching of scales, the men display a sense of identification with the product: it is 'their' beer. Benjamin argued that people often fantasize about potential connections between themselves and the dream world of advertising and commodities which are gathered together in the city streets, a fantasy which is heightened because the objects and experiences remain physically distant and unattainable (Buck-Morss 1991). This particular object, a symbol which reminds the friends of their good times together, provides a catalyst for play as simulation. A similar role is often played by public sculptures, many of which are also located at street intersections (see Chapter 9).

This advertising sign is large because it was designed to be legible from city buildings to the north and by oncoming traffic on Queensbridge Street. The tourists' perspective on the sign is incidental to its purposeful orientation, but it is this incidental exposure which makes a playful encounter possible. In this instance of distraction, an intersection is not experienced as a crossing of paths of motion, but rather as a crossing between people's line of movement and an object which suddenly becomes visible on a perpendicular sightline.

The tourists' photograph with the brewer's logo is a rare case of play at this intersection. This is somewhat surprising, considering the high pedestrian volumes, the surfeit of open space, good solar orientation, the proximity of numerous other leisure facilities, and the particularly long cycle of this crossing signal. One reason people do not stop and play here is that it is less pleasant than the remainder of the riverfront promenade. Four busy traffic lanes make the space noisy and polluted. The intersection is also very large, flat and windy. Because of the wide vehicular streets curving around the site, underground parking ramps, the cut-off end of a disused railway bridge and the porte cochere setback of the casino, there are no other activity spaces or occupied buildings within one hundred meters either side of this crossing point. The scale frames a great contrast between the experiential opportunities available to the man with the signboard on Swanston Walk and the three tourists crossing Queensbridge Street.

It is also significant that pedestrians move along only one axis through this open space. Foot traffic along Queensbridge Street is virtually non-existent, because of its range of land uses, large building scale and generally low amenity. This is only a stopping point along a pedestrian axis, and not a pedestrian intersection. Everyone stops or moves at the same time and everyone stands facing the same way, and so this intersection frames few chance encounters. Apart from the other pedestrians crowded together on the edges of this footpath, there are no different activities, moving or stationary, within the 20m perceptual field of the resting walker (Hall 1966),which might inspire people to act, or triangulate to start a social relation.

Comparing a large sample of play activities observed on Southbank Promenade and Swanston Walk, the most striking difference is that the great majority of play on Swanston Walk occurs at its intersections with other busy streets, while on Southbank play predominantly occurs on the long stretches of riverfront promenade between the intersections. Swanston Walk has a wide footpath with at-grade intersections every 100m. Many of its cross-streets also carry large numbers of pedestrians; two are car-free (Bourke Street Mall, and Little Collins Street at lunchtimes) and others have limited car traffic because they are narrow one-way lanes. Many performers, like the man with the signboard, site themselves at these intersections, because these are '100% locations' where they can expose themselves to the largest and most diverse audiences (Whyte 1988).

The four vehicular streets intersecting Southbank Promenade all carry large volumes of cars at relatively high speeds. These are wide streets with narrow, exposed, poorly furnished footpaths and in most cases poorly resolved connections to the Promenade. The high bridgehead of the southern continuation of Swanston Walk aids the smooth flow of people out of the city, and carries significant numbers of pedestrians, but leaves the road elevated high above the riverbank. It is connected to the promenade by two very steep, narrow stairways and an equally narrow ramp within a dark undercroft. Such links offer little scope for encounters with people heading elsewhere, for change of direction or for displays or lingering. Such slow points and junctions are natural stopping and meeting places; as such they should be of generous dimensions, whereas in fact they are far narrower than the promenades, as well as being very indirect and indistinct.

The intersection on Southbank Promenade where the most play occurred was at the south end of the pedestrian footbridge which connects to Flinders Street Railway Station and Elizabeth Street. This bridge has low springing points and opens onto the promenade at a wide junction which provides a natural social meeting point. Street performers often locate themselves immediately to the southwest of this junction in front of the Esso corporate headquarters (Figure 5.5). These displays are framed at the end of the axis of the bridge. People tend to slow down, look around and linger as they turn from the bridge onto the promenade, or as they negotiate the slight bend in the promenade path.

One busy intersection within Southbank which could conceivably be a site for spontaneous encounters and playful activity is a central interior foyer of the casino complex. This indoor space has many attributes of an urban intersection. It is framed by the entries to the casino gaming hall, the theme-driven All-Star Sports Café and restaurant 'Planet Hollywood' with

Figure 5.5 Playful performance fills the vacuum of a blank backdrop at an intersection: Esso building, Southbank Promenade.

a merchandise booth on its front corner, and the escalators down to the food court and entry on the ground level. Along one side of the foyer is the box-office of a twenty-four-hour multiplex cinema, and the lines of patrons stretch back into the circulation area, where others wait for friends, trying to decide upon a film.

One Friday evening as light rain falls outside, this foyer is busy with people seeking diversion, and is saturated with noise and moving images. Yet there is almost no spontaneous human interaction. A young crowd are spilling out into this space after a movie. One man in a large group crossing the foyer feigns running headlong toward the casino entry on the east side, 'drawn' to it, waving his arms to show his helplessness, with a huge goofy smile on his face. A friend pretends he is trying to drag him away, trying to talk him out of it. The man's behavior mimics the sense of abandon and distraction that urban space can bring. Everything brought together in this location is intense, fantastic, diverting. Patrons expect to forget their everyday roles and concerns, to lose themselves here. The man's playful act is one of *détournement*: through satire, he effectively dissimulates the contrived lure of the casino: pretending he is absorbed by the pretense, acknowledging its diversions but refusing them. Even though strangers are brought into close proximity in this space, it is actually very hard to look at other people. There is too much competition from manufactured distractions. The environment itself is so stimulating, no one seeks out the extra stimulation of strangers. Three sports-action video games have been placed around the entry to the All-Star Café. Loud, high-energy music spills over from a compact disc store behind the café. The noise level precludes overhearing other groups' conversations. Bright lights flash, adding to the perceptual excitation. A large video screen next to the casino entry shows prize winnings and delighted winners. Between the scrolling timetables above the box-office, movie trailers are showing. There are also collages of clips from other movies running outside 'Planet Hollywood'. Most of these television screens are set above eye level, a distraction which reduces the possibility of accidental eye contact between strangers, despite close proximity. Even the carpet, with its strong directional pattern that aligns the space with the gambling area, competes with the innumerable trajectories of feet swarming across it. Despite its representations of diversity and disorder, this foyer establishes an altogether controlled and passivized crowd.

Two very different intersections within Southbank, the promenade crossing of Queensbridge Street and the foyer deep within the casino, illustrate an extremely functional polarization of space within the overall design of this entertainment precinct. The former, outdoors, is an instrumental junction where only minimal pedestrian needs are supported. There is a lack of attention to the possibility of pedestrians lingering. The potential for sensory stimulation through experiences of intensity, distraction or chance encounter is limited by largeness of scale and the geometry of vehicular traffic, much of which services the casino complex. The indoor space, by contrast, saturates perception with carefully orchestrated sounds and images, at the expense of exposure to the truly unpredictable. One intersection optimizes physical circulation and dispersal, the other only gathers together images for passive reception.

This spatial segregation can be contrasted to intersections within the grid of the city, such as those along Swanston Walk. At these intersections, people are moving not just in different directions but with totally different purposes. Some of the images gathered together here are purely instrumental, others are escapist. It is possible to have close, intense engagement with a wide diversity of people and images at street intersections within the grids of the city. There is also the possibility of unexpected and unregulated intrusions and distractions, from new people and new phenomena. Urban intersections support both practical and playful possibilities, and integrate them together.

The dense interconnections of paths within the grid of the Melbourne CAD frame the possibility of unexpected encounters. The microgeographies of intersections give physical shape to people's encounters with urban difference as they move through the city. Pedestrians converge at intersections from different directions, and this impacts upon the prospects for any of them moving efficiently through public space, bringing confrontation, interruption and distraction. Observations show that the crossing of people's trajectories can, however, result in enjoyable, exploratory negotiations of their differences through play.

The majority of central Melbourne's urban fabric is built out to the property frontage. This keeps intersection spaces tightly framed, maximizing the suddenness of exposure to people and to new impressions. Many of the city's significant public buildings are located on corners, and this increases the amount of human activity gathered there. The man with the signboard and the cigarette artwork on the footpath show that the crowding together of people at intersections also increases the closeness of people's encounters with unfamiliar images.

The brewery logo is another example of an interesting view being suddenly revealed at an intersection and stimulating spontaneous play. The setting of Queensbridge Square illustrates some of the limitations which a wide, open intersection dedicated to cars has for playful pedestrian use. Southbank promenade in general shows how poor design of connections between paths can limit unexpected encounters, exposure to distractions and choices for movement.

Chapter 6

Boundaries

Boundaries set limits to what people can see, what they can do and where they can go. But in relation to play, which has diverse sources of stimulus and varied forms of conduct, boundaries also define many opportunities. Boundaries differentiate space. People make use of boundaries to shape their experience of the city and their play. Boundaries are popular places for people to spend their free time.

Physical boundaries help provide structure to social relations. They demarcate areas where people can perform different roles, as either audience or performer, determining their level of exposure and also different kinds of exposure: not all boundaries are equally solid. People's movements across boundaries raise or lower their level of playful involvement with others. Boundaries can define people's togetherness and their distinctiveness. Limits or barriers are also sometimes used to define people as different or to physically contain or exclude people. Such boundaries can become a catalyst for creative or transgressive behavior where people test the effectiveness of physical controls.

A second way boundaries within public space enable play is by defining a special place apart from the everyday life of the city. Such marginal places may be difficult to access and may be out of sight; they may have their own distinct social and environmental characteristics. This separation facilitates people's escape from functional activities and normal perceptions and from behavioral regulation by others. With such seclusion, people can indulge in new kinds of experiences and transgressive behavior. Examples of marginal play within riverfront leisure areas in Melbourne, Brisbane and London illustrate the general condition of such precincts as a place apart, on the margins of the city; a site for behavior which contrasts with serious, productive urban social life.

Physical boundaries within spaces are often explored and tested through people's play. Solid edges provide something to push or move against, they generate a tension which is often experienced through the body as vertigo.

The edge effect

The edge of public space is a popular place for people to situate themselves during their leisure time, because it offers protection, while allowing controlled exposure to outside stimuli. De Jonge (1967) calls this the 'edge effect'. People utilize edges and boundaries within spaces to regulate their level and type of engagement with strangers and establish a comfortable balance. People who are elderly, inexperienced or shy are often to be found on the periphery, because for them the public realm presents a wide variety of uncertainties, threats and hazards. Triangulation can result when people are watching play from the boundary of public space; they feel happy and relaxed there, and sometimes end up engaging with other strangers in the audience. Well-defined boundaries can also help contribute to people's sense of togetherness with others who are also inside them.

Boundaries limit visibility, communication, contact and movement; they restrict one's experience of other people. For such reasons, the edge of the public realm is also a more secure place to engage in playful activities. People play discreetly on the boundary of public spaces because there is less likelihood they will be noticed by strangers or compelled to interact. People pursuing clandestine, proscribed play activities also often take advantage of the seclusion of edges.

Figure 6.1 Passers-by watch play from the far edge of the street: Bourke Street, Melbourne.

People situated passively on the edge of public spaces sometimes engage with others around them as a result of sharing an unexpected entertaining experience. Office workers on their lunch-break smile and comment to each other as a woman confronts another driver who has boxed her car into a parking space. The chance event requires a highly theatrical negotiation between these strangers. The opposite footpath also becomes very congested with people stopping, watching and laughing (Figure 6.1). The motionlessness of the crowd on the footpath and the orientation of their gaze alert other passers-by to this event. On Swanston Walk, two men stand at a boundary of the stage which a street performer has established on the footpath, and talk about his performance (Figure 6.2). A third stranger sitting on the bench listens in on their conversation. Two strangers sitting on a bench outside the Post Office, their bags between them, begin a conversation about an improvisational comedian.

These are examples of what Whyte (1988) calls triangulation. The conversations center around the play of others, which in each of these examples is some kind of performance. The conversations themselves are playlike: unplanned, informal and non-instrumental. Conversation with a stranger is seldom the primary reason someone has come to a public space. The very

Figure 6.2 Performance close to the footpath edge (left) and triangulation from the safety of the edge (right): Swanston Walk, Melbourne.

unexpectedness of such an interaction makes it an escape from one's instrumental tasks. Even the most choreographed play event in public spaces intensifies both the possibility of such encounters (by concentrating people around its edge) and the atmosphere of relaxed abandon and curiosity which stimulates them.

The examples suggest a number of conditions associated with the edges of public spaces which facilitate triangulation. The first is that the inter-actions between the onlookers depend upon them having a view onto the stage of public space, where the events unfold. The edge provides the fullest overview of public life. Being at a distance from the action, for example on the opposite footpath (Figure 6.1), means a wider field of view. Triangulation is also likely at the edge because people are generally at rest there; they can pass time comfortably. The most out-of-the-ordinary events in public spaces are spontaneous and fleeting, and thus chance exposure to them is more likely when people spend more time watching. Planters, window ledges and steps in front of buildings are locations where people often sit and watch others play. In Melbourne, the predominance of verandahs over footpaths means that the edges of spaces also provide shelter from sun and rain. The three events mentioned all occurred at liminal times, lunch or evening, when

people could linger if something interesting happened. People appear to spend a great deal of their free time at the boundaries of public space.

Being at the edge limits the risk that people will be disturbed from their watching. People walking in public space who want to stop and focus their attention on some point of interest typically back up to a boundary to avoid collisions with others. They seek out pockets of relative passivity which do not function as transit areas and where they are not themselves on display (Figure 6.3). Comfort includes the psychological comfort of remaining discreet. Play is not enjoyable if one is forced into a high level of involvement. The boundaries where people can be seen watching play in public spaces are often smaller scale, darker and less trafficable. They allow seeing without being singled out as an observer. Most benches in the Bourke Street Mall are laid out backing onto planters or flagpoles or each other. This allows a large number of people to sit at a 'boundary' and watch those performing out in the open areas of the Mall, without being apprehensive about sudden encounters from behind.

Triangulation is also more likely at the edge of public space because the attention of people watching from an edge is almost always focused in just one direction. This orientation makes even the close physical presence of many other strangers at the edge tolerable. Under such conditions it is easier and less confrontational to begin and to break off conversation.

Figure 6.3 Crowd viewing solar eclipse from the safety of the margin: in front of Galeria Kaufhof department store, Alexanderplatz, Berlin, as well as from balcony (center top).

These observations refine the definition of the edge effect in relation to the specific dynamics of playful encounter through triangulation. The human desires for refuge and for exposure to the world are in a tension which is mediated within specific physical settings. Privacy and security is weighed against the desire to gain the attention of others, to be within a distance at which they can be addressed, the desire to escape from practical thought and action and experience a world of new sensations. The cases described above show people choosing the risks of exposure without clear instrumental benefit. Contact with strangers is a part of their enjoyment of being in urban public space. Triangulation remains a pleasurable experience when the individual retains a sufficient measure of control over the circumstances of their exposure. Edge spaces, being between two realms, embracing both open and closed orientations, tend to offer such control.

Both sociofugal and sociopetal seating are defined for specific levels of social engagement (Hall 1966). Movable furniture is not often provided in public settings, and the high level of user control it allows suggests that it makes chance interactions less likely (Whyte 1980). Straight rows of continuous seating, such as long steps, maintain a small level of exposure risk which people can modify through their own body language, by the use of props or by adjusting the distance or elevation that separates them.

Remaining at the boundary does not mean total passivity. Because the edge of public space is more discreet and better protected from incursions, playful acts by members of the public also often begin at the boundary. Generally 'activities grow from the edge toward the middle', as people move from a defensible zone into other settings which are more open and risky (Gehl 1987: 152). Play situations begin in a similar context, as an individual steps beyond their comfort zone toward the unknown.

People often play games on the edge of public space. Two examples are people playing chess (Figure 6.4) and backgammon on the footpaths outside cafés. The games themselves frame an escape into a world with rigid rules of encounter. But the locations in which these people play suggest that their desire for escape is not total. The man at the rear table in Figure 6.4 speaks to a friend who has chanced to pass by. Exposure to people and events outside the gameboard also remains a possibility for the couple playing chess. The chess game, like the tables and chairs which frame it, provides a way for the couple to comfortably and pleasurably pass time on the edge of the public realm. A similar spatial arrangement frames two men playing backgammon at an outdoor café table on Lonsdale Street. One player slaps a counter down loudly and emphatically as he moves it, making it clear that their performance in the game is subject to public scrutiny. This location has a waist-high rail giving the players protection on three sides against passing cars and pedestrians. This arrangement allows pedestrians to more comfortably view the action at close proximity. A partial boundary provides security without completely preventing exposure.

Figure 6.4 Edge as secure space for play: Little Collins Street

Each of these instances shows people enjoying a leisure activity for its own sake. But to varying degrees, each is also a simulative or competitive display at the edge of public space. The level of exposure of the players is shaped by many environmental factors: the daylighting conditions at the edge, the line of sight of passing pedestrians, boundary markers, barriers and changes in elevation all define a place apart for their play.

Adjacency

Because a boundary is a limit to the movement of people through public space, it can also act as a backdrop for playful displays, enhancing visibility while optimizing the performer's control over their own engagement with the spectacle. A wide range of urban design elements in Melbourne's public spaces define boundaries which frame staging areas.

Most street performers in Melbourne are to be found along the edges of the city's main pedestrian paths, the Bourke Street Mall and Swanston Walk (Figure 6.2). Performers need an audience; they must use a space which is adjacent to existing flows of people. They have to be able to distract people from their business, and in most cases this is done by capturing the gaze. Every performance has a front, and passers-by need to be able to see this front from where they are already walking.

The performers set up to one side of the path. This allows them a stage area where they can move around. The line of vision between the performer and any stationary onlookers is generally perpendicular to the main pedestrian path, so that it does not actually obstruct motion. In the Mall and Swanston Walk, the width of the footpath also accommodates an edge space out of the flow where the audience can stop and stand. People who stop and watch performers on the wide pedestrianized Mall are free to stand where they choose, but the performers distance themselves within the physical context to frame a certain relation. The majority of the audience remain approximately 6m away from the actors.

Hall's (1966) work on the dynamics of social distance provides useful yardsticks for understanding why public performances occur where they do. Between 5m and 6m is a common distance for theatrical performance. At this separation, the whole face of the performer can be seen in sharp focus. Within a 60-degree cone of normal vision, the entire bodies of performers can be viewed, along with some of the space surrounding them. The statue act relies on successfully disguising the humanity of the figure by painting the face, hands and clothes as if they were all made from the same material, and holding static poses. It is thus to the benefit of this deception that from a 6m distance, 'fine details of the skin and eyes are no longer visible' (Hall 1966: 117). However, at 6m separation, facial expressions and minor body movements remain discernible; the actors do not have to exaggerate their actions. This potential for subtlety is a crucial element of the performance, which focuses on a tension between the absolute stillness of a statue and the graceful motions of the actors when they suddenly, unexpectedly give a genteel nod or wave.

Although there is enough proximity for the audience to maintain a studied view of the actors, engagement at 6m remains formally structured. The distance is well outside the possibility of physical contact. If a passer-by wants to approach the statues to see whether they are real, or to touch them, they have to move inside the tacit threshold established by the rest of the audience. People approach warily, unsure of just how frozen the statues are or what they might do, because at this distance the statues have time enough to take evasive action against any approach.

Members of the public often resist street performers' high-pressure entreaties to become involved in their events. People are not always ready to respond to play opportunities which suddenly, unexpectedly present themselves. They generally choose to watch or participate only if they feel they are doing so freely. Performers can distract and cajole to keep an audience's attention, but ultimately they have to arouse a desire within the passer-by to stop, relax and be entertained. Crowds gather and stay under circumstances where individuals retain some ability to adjust their level of involvement by moving back or forth, by sitting (becoming more passive) or standing. Some people like the opportunity to view a performance discreetly from a great distance.

The width of the Bourke Street Mall allows performers to stand far enough back from the flow of the footpath that pedestrians, if distracted by the performance, can feel comfortable stopping on the spot to watch. Involvement requires very little physical effort. This distancing is also possible on Swanston Walk. However, the statue in Figure 6.2 stands much closer to the footpath, raising the level of involvement. At a separation of 2m, if passing pedestrians turn to engage him, they are already close enough to initiate bodily contact. This statue shakes hands quite often. The intended audience can bear such proximity because they experience it only peripherally and momentarily in passing. Those who want to watch him for longer, to see if and how he moves, actually move away from the performer, back to the opposite edge of the footpath, in order to avoid conflict with people walking past.

This finding has certain implications for the design of pedestrian ways: purely visual engagement seems to be most comfortable where there is a width of about 6m between the two edge zones where the actor and the audience can securely position themselves. Footpaths and arcades with lesser widths frame a higher level of stimulation. This will either lead to more direct, bodily forms of interaction, or necessitate different kinds of coping mechanisms, such as civil inattention (Goffman 1980).

The footpaths of both the Bourke Street Mall and Swanston Walk were widened when traffic was removed from these streets, creating intermediary edge zones free from both vehicles and pedestrians. Figure 6.2 shows that performers on the outside edge of these footpaths are made conspicuous because they stand in bright sunlight, outside the shadow of the edge of verandahs which lined up with the original street kerb. These excess zones along the street edge have been partially filled with street furnishings such as benches, telephones, information panels, garden beds, outdoor café tables and kiosks. The discontinuity of these linear zones prevents their use for the function of movement: they are places set apart from instrumentality. Street performers, chalk artists and musicians are just another superfluous, luxurious appropriation of this 'lag' zone at the edge. It is because opportunities for playful experiences of escape lie available immediately on the edge of the urban footpath that walking can easily become something other than instrumental.

Audiences are more likely to gather when such edge zones of public spaces are conducive to lingering. Alexander et al. (1977) argue that lingering is a gradual process, whereby people moving through a space are distracted from their instrumental tasks. If places for lingering and places that support protracted activities such as performances are physically adjacent to paths where people pass, the 'goal-oriented activity of coming and going then has a chance to turn gradually into something more relaxed' (Alexander et al. 1977: 600). It is only under relaxed conditions that play becomes possible.

A particularly comfortable and relaxed boundary condition is presented at Brisbane's Southbank leisure precinct, where a large free public swimming pool with an artificial beach lies adjacent to a wide riverside promenade which permits cycling and skating (Figure 6.5). The pool has been built to enable eye-level interaction between swimmers and passers-by, and steps lead up from the promenade to a wide sittable edge. The edge of the pool has been designed for easy cross-visibility and cross-circulation between zones, an interweaving of leisure activities, while ensuring adequate segregation of the very different experiential realms. Swimmers, sunbathers and strollers are all on display to each other. Careful treatment of heights allows pedestrians to very easily touch the water. Brisbane has a warm subtropical climate, and the tantalizing view and the ease of crossing this boundary often encourage people to spontaneously take a dip in their clothes. Swimmers can also easily climb out onto the promenade.

Backdrops

While street performers select locations where they can easily distract passers-by, they also need ways of shutting out external distractions. They want people to forget about their predetermined goals and their urgent responsibilities and be captivated by the show. One technique is to choose

Figure 6.5 Interface between riverfront promenade and edge of public swimming pool: Southbank, Brisbane.

a blank backdrop. An example is street performers in front of the Esso corporate offices on Southbank (see Figure 5.5). Here they can easily capture the attention of people who look their way, because the building frontage behind is inaccessible from Southbank Promenade. There are no café tables, no signs, nothing to consume; just a featureless glass curtain wall fronted by a low flight of steps. A performer can stand here with sunlight falling on them from the north, and easily become the center of attention. Further along in front of the equally blank waterfront offices of Philip Morris, musicians and gold- and silver-painted 'statues' line the promenade, part of which is usually covered by the chalk of pavement artists. The absence of cross-streets on Southbank means limited opportunities for chance encounters, but this condition is advantageous for staged displays, because it creates a closed edge, reducing the possibility of interruptions. It orients space with a clear front and back. In this setting the actors can easily 'face' the flow of pedestrians. It is much more difficult for an actor to perform 'in the round', where the boundaries of front and back stage and audience position are not so clearly defined.

Blank backdrops are not the only kind which contribute to a scene. Performances also happen on the opposite, river edge of Southbank Promenade. Thousands of pedestrians pass along this footpath daily. Two teenagers ride around on an open area of paving, doing a wide range of skilful cycle tricks: spins, flips, wheel stands, jumps. They get the attention of a lot of the passers-by (Figure 6.6). There are many other sites in the city where these cyclists could try out a much wider range of more thrilling tricks: this location offers only one kind of low, flat bench to jump on. The most appealing features of playing here are the passing crowd and the backdrop.

When pedestrians come abreast of them, these cyclists are still a far public distance of 8m to 12m from their spectators (Hall 1966). Over such a distance, only exaggerated actions such as theirs can readily capture the attention of passers-by. The cyclists often hold frozen poses such as front wheel stands so the passing crowd gets a good look. This is a distance which also allows the audience to easily follow the rapid movement of the players. The distance of 8m to 12m defines a certain relation between actors and audience. In the case of the human 'statues', proximity was predicated on the tension of increasingly close examination to test the deception of the act, the surprise of occasional hand contact with passers-by, and the practical importance of passers-by being able to place money in the performer's hat without having to veer too far from their course. With the trick cyclists, there is no expectation of physical or financial engagement. In fact, proximity would bring undesirable risks, and they would have to curtail their exuberance. These performers do not want money, merely recognition, and distance helps maintain the aura which their display of skill establishes. They are content for the audience to remain uninvolved bystanders. The great width of this edge of the promenade thus allows these players a suitable combination of separation and visibility.

Where these cyclists perform, the promenade bends in an 'S' around a restaurant which projects from the casino and then back in the opposite direction to meet the intersection of Queensbridge Street (see Figure 5.4). This frames the cyclists in the direct line of sight of pedestrians approaching them from both directions. Their stage is framed as a dead-end 'pocket' by an uneven plot of grass and a series of large columns which spout water and fire. On the inland side of the promenade, a thick hedge screens the casino's hotel driveway. The gaze of pedestrians walking along here tends to stray only toward the expansive view north across the river. For those walking east away from the casino, the slight curve of the riverbank also places the city skyline as a stunning backdrop beyond the river, dramatically lit by afternoon sun. Many promenaders are inevitably already looking in the cyclists' general direction, and they interpose themselves on the spectacle. The distance of 8m to 12m allows the cyclists to be viewed within this wider setting. They locate themselves between the pedestrians and this visually stimulating backdrop which emphasizes their own urban credentials.

The high visibility of this stage on the edge poses a challenge to the efforts of the casino management, who use quality urban design and a fabricated atmosphere of permanent carnival to draw patrons from the city center along the lengthy river axis and into the casino complex. Informal acts of play such as these cyclists thrive on their adjacency to the crowds and excitement which the casino can marshal, as well as the surfeit of unprogrammed open space provided here. In an interesting dialectical development, the casino management subsequently eradicates the potential for distraction by such behavior. They tighten the promenade space, plugging the wide, open boundary area with new planters and sculptures which block the cyclists' orbit (Figure 6.7). The form and orientation of the new sculptures show a conspicuous effort to redirect the attention of promenaders along the major axis which leads to the casino, recapturing the idle gaze of tourists for more profitable forms of escapism.

Both the blank corporate office facades and the 'unfunctional' excess width of the Southbank Promenade are contrary to conventional urban design wisdom. Poor design has a certain benefit in the particular context of this spectacular leisure landscape because it leaves room for new, playful contributions to the diversity of the urban scene. But at least three conditions are necessary for such empty boundary areas to have a chance of stimulating playful performances like these. First, the boundary areas have to be located along a route which is already frequented by a potential audience. Such unexpected distractions are merely complementary to the primary leisure attractions which are spaced along this riverfront (see Chapter 5). Permanent, spectacular displays of spontaneity, exuberance and release on the promenade are instrumental attempts to stimulate unrestrained consumption, with the public generally framed in a passive role as spectators. In between the sudden fireballs, leaping fountain and colorful public artworks along the promenade's edges, there are gaps where more informal, participatory

Figure 6.6 Performance at a distance: trick cyclists, Casino Promenade, Melbourne.

Figure 6.7 Spectacle fills superfluous space: Casino Promenade.

performances can take place. Second, the boundary spaces have to be both adjacent and highly visible. Third, the management regime has to be either permissive enough or lax enough for actors to be able to appropriate these sites. The casino's ongoing efforts to mend gaps in the spectacle show that the struggle between the management of leisure scenography and the public's free play continues to evolve.

Control and transgression

Performers use the edges of space to provide backdrops to their display and to articulate distances and orientations between themselves and their audience. They also attempt to establish and maintain various kinds of boundaries between their stage area and their audience, so that they can maintain adequate control over their performances.

People playing on the edge of public space sometimes make use of windows, glass barriers which allow visibility but also security because they prevent direct contact with audiences. In Melbourne's Albert Coates Lane, three children imitate the posing of mannequins – and each other – in a fashion store display window, to the amusement of passers-by (Figure 6.8). The window eliminates the risks of being approached by their audience and of their deception being uncovered. It allows onlookers to come much closer than would otherwise be manageable, heightening the thrill of the act, as the children struggle to suppress giggling.

This window theatre also confronts and transforms a general expectation about the privately managed edges of public space. Such windows are commonly used for artful displays of goods which aim to distract the attention of passers-by, arouse their desires and stimulate impulse purchases (Crawford 1992). The children's performance draws upon the same power to distract, but subverts its instrumentality, substituting the momentary pleasure of consumption with the exploratory, difficult and unpredictable pleasures of performance. The children turn the window into something active. They do so unexpectedly and temporarily, and as Bakhtin noted of festivals in general, 'the very brevity of this freedom increased its fantastic nature' (Bakhtin 1984: 89).

Not all forms of boundary are as secure as a display window, and audiences are often too far away rather than too close. The boundary conditions within urban space provide a frame of possibilities which performers usually have to reinforce. Performers use a variety of behavioral techniques to sharpen these boundaries. In many cases, street performers draw a chalk line on the ground at a 6m radius, and repeatedly ask those who are watching to step up to it, and also to keep a space clear behind for pedestrians and vehicles that need to pass. This ploy increases people's level of involvement with the act, and thus reduces the risk they will be distracted and lose interest. The relation between audience and performer is itself a game which keeps changing. In the cases of play considered below, members of the public at play make incursions upon the spaces of strangers around them. These examples focus on people transgressing social boundaries as they change their level of playful engagement with strangers.

The typical 5m to 6m separation between street performers and their audiences can be seen in an example of public play in front of the Myer Department Store (Figure 6.9). Where people choose to form the edge for viewing depends in part on the size of the audience and the feasibility of blocking the flows of pedestrians moving along the footpath and in and out of the store. Pedestrians pass by along the footpath and sometimes stop on the near edge to watch. Those watching for longer tend to occupy the areas at each end of the zone, particularly the benches at the east end which face onto it. The subtle use of lighter stripes in the paving pattern seems effective in suggesting an appropriate boundary for watching. The arrangement of street furniture such as light poles, planters, telephone booths and bollards a short distance back from this implied edge provides a safe backdrop against which onlookers can stand, without having to be concerned about being in the way of pedestrians. Together the paving and street furniture delineate an edge zone which is also useful for the performers. These urban design

Figure 6.8 Children pose as mannequins in shop window to the amusement of passers-by: Albert Coates Lane, Melbourne.

elements give structure to the play of performance, without preventing people from choosing their own level of involvement, and without preventing through-passage by disinterested pedestrians, which may lead them into spontaneous involvement with the performance.

This example shows two elderly men and three young female tourists who have stepped forward from the 6m audience edge to dance to the music of a sitar player. They dance together, sometimes holding hands in a circle, sometimes twirling in pairs. They learn each other's dance steps by imitation. The social practice of dancing frames special, regulated forms of close encounter between these strangers. People standing and listening to a performer need some excuse to step toward and interact with others occupying the public space around them. Dancing provides that excuse. The edge established by the audience provides a space apart from pedestrian activity which the dancers can make use of. They can become absorbed in the music and freely twirl. To an extent they upstage the musicians: their play distracts from the more instrumental performance.

As a field of play, the Bourke Street Mall is a complex overlapping network of many sub-spaces, each of which is subject to definition and defense by its users, and to observation and intrusion by others. There are prominent

Figure 6.9 Audience acknowledges a boundary: street musicians and members of the public dancing, Bourke Street Mall.

spots for people to perform, a series of relatively secure, discreet locations for others to watch in comfort and consider options for encounter, and an intervening series of distances or marked boundaries which can be crossed by the movement of either party. Built elements such as benches, planters, footpaths, tram tracks, paving patterns and the shadow-line of verandahs all frame this field of encounter. People make use of this spatial field to approach others and heighten their experience of other life possibilities, their experience of escape, while also protecting themselves against an unwanted level of exposure. Both angle of orientation and distance are important in defining the level of engagement between parties. The sociopetal or sociofugal organization of benches in a public space like the Mall appears to have a considerable impact upon where people feel comfortable sitting and how they engage one another. Ideally settings for playful encounter provide opportunities for people to have close exposure to a performer without necessitating commitment. This can occur with performances right next to a footpath: initial encounter is only peripheral and fleeting. Such close exposure can stimulate and encourage play.

The spontaneous nature of play, the diversity of people's desires, and the implied freedom of public space mean that members of the public do not always respect the boundaries for seeing and being seen which are laid down by street performers. On London's Leicester Square, now no longer open to performers, a group of street musicians play African drums to an inter-national audience of onlookers. Their rhythms inspire a pair of Spanish tourists to step out onto the small stage formed by the tight-packed crowd and give a skilful and exciting display of Latin dancing (Figure 6.10). Their example prompts others to step forward and dance, which ultimately provides a basis for encounters among dancers. The smiles of the Spaniards and the trepidation of other onlookers who step forward highlight the significance of this practice where one communicates a lot about one's self, one's body and one's culture. Here in Leicester Square, a cinema, theatre and nightclub precinct where culture is being presented for consumption, people can also cross the line to produce social identity. Informal, spon-taneous, playful performances maintain an important role in igniting the public imagination.

Breitscheidplatz in the heart of the former West Berlin is a similar point of congestion and high exposure. There is a constant tension here from the crossing of lines of movement and lines of vision. Tourists, local citizens and performers find it difficult to structure their interactions in space. A group of street performers set up where the bending of Kurfürstendamm frames them right in front of westbound pedestrians and where the wall of the Kaiser Wilhelm Memorial Church forms a blank backdrop. They try to arrange their audience, to secure this provisional stage. But as the show begins, two of the plaza's regulars allow their dogs to start a noisy fight on the raised terrace behind. The dancers are literally upstaged, and their leader has to go to negotiate with the owners.

Figure 6.10 Spanish tourists emerge from the audience and start dancing in response to African drummers: Leicester Square, London.

Tensions between watching a performance and the desire to play can also develop among members of the public. A band plays rock classics on Breitscheidplatz for donations. Several people dance on the spot, but one woman really lets loose: she swings her arms wildly, jumps around and does exuberant high-kicks. She is a big part of the entertainment. A few songs later she and a nearby stranger have a heated argument. Her flailing arms are apparently getting in the way of his video recording of the band. The friction is between willful abandon to the effects of the music, and its passive reception as a spectacle. The public is not a unity; their desires for pleasure are varied.

The indeterminacy of spatial territories within urban public space often generates tension. It highlights the need for constant boundary maintenance by all of its users. But boundary maintenance is not just inspired by fear and mistrust. It includes actions of approach to others, such as surreptitious observation while walking past, or actions which invite others closer. The effect of urban space is, after all, to bring people together in a shared space, not to segregate them. Some interactions between people are non-instrumental; they are without consequentiality. Hence the tension can be great but the risks are minimal. There is uncertainty, because roles, rules and relations are not well understood or mapped out on the ground, but there is not necessarily danger. The conditions of publicity and anonymity which prevail in public spaces allow strangers to approach each other more closely and engage more freely.

In another example of play that transgresses a socially defined boundary, the act of bounding and policing space itself stimulates an exuberant playful response. During the street parade of Melbourne's annual Moomba festival, marshals walk up and down asking the crowd to stand back behind a blue safety line laid down along the edge of Swanston Walk. To bring an element of fun to the pragmatic task of segregating crowd and performance, the organizers have created a huge foam figure, a corporate executive with a bulldozer instead of legs (Figures 6.11 and 6.12). Apparently very pressed for time, he rushes around, talking on his mobile phone and thinking of money, buffeting people to the side. Most people toe the line. But the ideal location to view the parade and to get the maximum sensory stimulation is as close as possible. Separation produces a tension. The procession intensifies and enhances people's desire, but the boundary line deprives the audience of fully satisfying that desire. Crowding, noise and the anticipation of the event generate a sense of vertigo. Some people react spontaneously and

Figure 6.11 Fun way to make people toe the line: Moomba parade, Swanston Walk, Melbourne.

creatively to the figure's efforts, inventing games based around him and what he is trying to do. They step over the edge, try to get their photo taken next to him, shake hands with him, pull his nose, or run away dodging him. They act against his practicality. This illustrates how the experience of escape through play is defined oppositionally in terms of what people are not supposed to do. In this latter example, the social boundary of a public play space serves to exclude the public from a role as players, but people struggle against the spatial organization of the parade as a passive consumption

Figure 6.12 People resist passivization: Moomba parade, Swanston Walk.

experience and the restriction of behavioral excess at the Moomba festival. Their games evoke Bakhtin's depiction of carnival in the Middle Ages as a ritualized rejection of social norms of prudence and propriety, and as an event without boundaries or limits:

> carnival does not know footlights, in the sense that it does not acknowledge any distinction between actors and spectators . . . Carnival is not a spectacle seen by people: they live in it, and everyone participates because its very idea embraces all the people.
>
> (Bakhtin 1984: 7)

A further example shows people's willingness to engage playfully with a physical boundary which restricts their movement within public space. The Pancake Parlour chain restaurant has set out two rows of poles along one footpath of the Bourke Street Mall, leading from Swanston Walk approximately 50m to the escalator at the entry to their business (Figure 6.13). Helium balloons printed with their logo are tied to the poles with string. There is no string between the poles, so the barrier does not actually prevent people cutting across the passage. But the partitioning of the footpath challenges and structures people's normal freedom of movement. Pedestrians are apprehensive when they approach the start of the rows. Perhaps they will have to do something if they go along between them, perhaps the poles are linked. One man tells his friends: 'It leads you all the way to the Pancake Parlour'. Many people, particularly those older and less able-bodied, detour around the outside of the rows to avoid potential problems, minimizing their risk of the unknown.

The setting creates a sense of vertigo. It compresses the pedestrians into less than half the normal footpath width, pushing them up against the dense lines of colorful, drifting balloons. Oncoming strangers are also brought up close. Sometimes it is very windy; the balloons blow horizontal across the walkway. This leaves less than 1m clear to walk, and people have to yield to oncoming pedestrians.

Some people contest this bounding of public space. They walk straight across the walkway, against the flow suggested by the lines of poles. An older woman heading across from the Mall space toward the storefronts pulls on a string to drag a balloon with her, disturbing the line. The balloons inspire a little competition, in the willful, destructive mode of *paidia*. The woman resists being directed during one of the few times and in one of the few places when she can act freely. Some pedestrians even cut diagonally across the rows. Rather than consciously contesting, they are completely ignoring the boundaries the balloons suggest.

Pedestrians develop an amazing range of playful ways to respond to this intervention. A woman tugs on the strings as she walks past (Figure 6.13). Several other people run their hands across the taut strings, playing them like a harp. A conservative-looking man in his thirties, walking with his wife,

Figure 6.13 Fun which regulates public space: people play with limits, Bourke Street Mall.

head-butts one balloon. Another younger man weaves in and out between the poles, lightly head-butting every balloon. A businessman with a tie and briefcase inconspicuously swats balloons aside with the back of his hand. An elderly woman waiting at the west end with her family, after having walked the length, goes back to punch some balloons just for fun. Most of these responses are competitive: people turn the stricture to their own advantage, explore its potential, making it serve their own needs and desires. There is one instance of simulation: the setting encourages a young woman to softly punch each balloon down and forward as she passes. Her example inspires a man walking several paces behind her to try the same thing. This is evidence that people pay attention to the games being played by strangers around them in public space. The different techniques which people use to displace the balloons also point to an element of vertigo: this is an encounter with an unknown physical property, the mass and inertia of the balloons.

Two further instances show a more pronounced tendency to vertigo, both physical and psychological. One man pops a balloon with a lit cigarette and laughs to his friends, one of whom then duplicates this act. Another man wearing a malevolent, gleeful smile is popping each balloon with something sharp as he passes. These individuals lose themselves in sensory fascination and shock at the sudden loud explosions. The act is a harmless expression of desires for violence and destruction. By making a personal mark on their physical environment, they realize a sense of agency in the public realm. The boundary thus provides an opportunity for display.

But this playful act can also be seen as a more specific critique of the social space of the balloon corridor. Here, as in the Moomba parade, a boundary which regulates people's use of space for instrumental reasons is dressed up as an opportunity for playful escapism. Finding play in the destruction of this boundary contests not just its physical constraint, but the symbolic packaging which suggests that voluntary submission to the boundary is fun, and therefore desirable. This vertiginous behavior is thus also a reaction to a behavioral conformism which is only implied by this boundary condition. Through the sudden release of air under pressure, these men burst the bubble on a business's imposition of a confined notion of play.

The loud pops of bursting balloons draw a lot of attention. An older man in a work uniform chides the pair from a distance, telling a workmate that he knew this would happen. The men's transgression emphasizes that the effectiveness of such a boundary within public space depends on its constant reinforcement through social action (Bourdieu 1977). It relies upon acceptance by an audience. Hence it remains open to playful transgression and reinterpretation.

When a protest meeting takes place on the steps of the Post Office, the council partially encloses it using a temporary barricade. The intention of this boundary is to stop this event interfering with the movement of trams along the Bourke Street Mall. However, it also serves to gather the crowd close, heightening the participants' perceptual contact with each other, their sense of taking part. Some participants also see this boundary as a convenient, secure place to rest and take in the scene and to park bicycles out of the way. The provision of a defined edge organizes activity, making the space more functional for them as well as the trams. Figure 6.14 shows some of the attendees playfully using the barrier to display symbols which illustrate their cause. Through their action, a boundary which is supposed to restrict interaction between the protesters and other functions in the Mall becomes a means of increasing communication with those who watch the protest from outside the barrier. A limit is turned into an opportunity.

The boundaries which differentiate between locations for playful performers and audience in public space can have benefits for both parties. They ensure the success of play in the form of display, allowing players improved visibility and room to move. Creating limits is not necessarily about excluding other people: keeping people out can also function as a way of drawing them in. The spatialization of roles within a public setting relies not just on physical interventions, but also on negotiated agreements regarding what role those interventions themselves play in social action. Members of the public retain control over how they interpret and respond to social boundaries. Where the risks are low and the stimuli are great, transgressions of these boundaries can take playful forms. Such transgressions do not necessarily destroy play, as Huizinga (1970) suggested; they merely change its rules.

Marginality

The cultural and leisure precinct of Melbourne Southbank is a boundary zone which lies at the margin of both the CAD and the Yarra River (see Figure 4.1). Performances on Southbank draw upon the density and diversity of people who live, work and shop in the inner city. The edge zone of Southbank can easily be reached on foot from the Central Activities District. Yet its physical and social character is quite distinct. It is separated from the noise and car traffic which keep pedestrians tense and alert. It has high environmental quality, facing north toward the sun and sheltered from wind, free from overshadowing by skyscrapers and from pollution (Gehl and City of Melbourne 1994). These conditions all arise through its separation from the city and from city functions: quality comes through control. Southbank is a marginal place (Shields 1991), adjacent to and yet disconnected from the psychological and perceptual intensity of the city center. It is a wide open zone of relatively unregulated public access and use.

The whole Southbank precinct is designed for leisure, but many kinds of playful, exploratory, transgressive, socially unacceptable and active forms

Figure 6.14 Limits to self-expression become an opportunity: Bourke Street Mall.

of escapism which can be observed here happen in spaces which are marginal to the leisure atmosphere of Southbank itself. The waterfront promenade is split, and a lower concourse, a discontinuous edge which cannot easily be seen from above, is often quite empty. Teenagers often hang out here, smoking, kissing or wrestling (Figures 6.15 and 6.16). This is a marginal peak-tide flood zone, and proximity to the water's edge adds to their wrestling contest the risk of being thrown into the water by or with one's opponent, of then having to combat the murky unknown. Teenagers also congregate on other more marginal walkways and darker corners at Southbank, using retaining walls, planters and the shade of trees to screen their activities, hiding away from the adult play-world of the busy main Southbank Promenade. For them, these are places set apart from adult surveillance and control. The kissing couple stand barely out of sight around a corner; there is an acute tension between their clandestine indulgence and the risk of being suddenly confronted by some prudish pedestrian. The proximity of the public helps stimulate the playful mood. Turned away from the promenading crowd, these are not displays for strangers, but rather a

Figure 6.15 Transgression in a marginal space: teenagers kissing, Southbank Promenade.

part of the identity formation of teenagers seeking to explore and extend their bodily potential through direct engagements with each other and with the material landscape. These mild acts of passion and violence hint at the escapism which a discreet but accessible urban space by the water's edge can offer (Jinnai 1995).

Another way in which the river stimulates play is that the sound and motion of the water and breezes stimulate the senses, distracting from instrumental concerns and awakening the body to action. The river's relentless motion expresses the raw power of nature. At the same time the water's movement is smooth, sensual; its flow conveys a sense of release. It is ephemeral, suggesting the spontaneity of play. The river provides a counterpoint to the rigidity of the adjacent urban environment. One evening during Melbourne's Formula One Grand Prix, two men stand on the edge of the river and gently, idly wave large Finnish flags back and forth in the strong breeze, watching them ripple and flutter (Figure 6.17). There are many people spread around the promenade and on a passing cruise boat, but these men are not showing off their interest in racing. They are in an almost solitary

Figure 6.16 Transgressive play in a space apart: teenagers wrestling, Southbank Promenade.

reverie, playing with the wind that blows unimpeded up the river. This playful act is framed by many oppositions. These men are fans of an extremely urban, global, expensive, fast and noisy technological contest of human skills. They stand right next to the busiest pedestrian space in the city, surrounded by a crowd of strangers and facing the downtown skyline as it comes alight. And yet their play shows them absorbed with an escape into simple, sympathetic engagement with a subtle, invisible natural force at the river's edge.

Similarly on Brisbane's Southbank, many people choose to sit on the low concrete barrier wall on the outside edge of the embankment, either cross-legged on top of it, or with their legs dangling 'outboard', to feel the breezes and to have an uninterrupted view of the river (Figure 6.18), even though this is not the most comfortable or safe option: plenty of comfortable seating is provided on the inland side of the broad promenade.

A last example highlights many of the main objectives and spatial conditions which characterize marginal play. Adjacent to London's South Bank promenade one summer evening, a couple cook dinner on a charcoal barbecue and toast each other with glasses of wine. Their choice of location is a tiny concrete landing projecting out from the embankment (Figure 6.19). The landing is fenced off from the busy tourist promenade directly in front

Figure 6.17 Exploring the edge of the river: Casino Promenade.

of the Tate Modern gallery, and there is a sign warning against trespass. On its opposite side the landing is completely unfenced from the 10m drop down to the water. Here the couple have a good view of the passing crowd, but are also alone on the river with an unexcelled and uninterrupted view across the water to the dome of St Paul's Cathedral.

A place apart

The kinds of activities described above could be pursued almost anywhere. However, the players have gone to considerable efforts to choose these particular marginal settings and to engage with their special qualities, and this choice is a significant aspect of what makes the activities playful. Four characteristics of experience stand out: having a view of the city from the outside; being close to crowds of other people at leisure; being separated from this crowd, even where crossing a boundary involves taking a risk; and engaging with water and the river breeze. The first two dimensions conform to the conventions of urban riverfront leisure redevelopment, whereas the last two tend to require contravening them.

In the history of city development, leisure activities have often been banished *extra muros* or have escaped there to secure their own freedom of operation (Kostof 1992). Leisure is, by its nature, a diverse, inclusive, messy

Figure 6.18 People sit with their legs dangling over a high embankment: Southbank, Brisbane.

Figure 6.19 A couple prepare a barbecue dinner on a precipitous landing beyond the safety railing of the promenade outside the Tate Modern gallery, South Bank, London.

'function' which does not sit easily with the desires for order and predictability which drive city planning and management. Leisure does not tend to respect boundaries. Particularly in protestant cultures, leisure has always been an uncertain virtue, best rigidly isolated from serious activity, although even the most licentious activities never move too far away from their customers. In Melbourne, Brisbane and London, the opposite riverbank has always been defined as the appropriate place for play, and in all three cases this now takes the form of a formalized cultural and leisure precinct named Southbank. The close physical juxtaposition of work and leisure contributes significantly to the experiential quality of these three Southbanks. Large numbers of people empty out of the cities' office precincts at weekday lunchtimes and evenings, and accordingly the vast majority of playful events in these precincts occur in these kinds of marginal time periods. The spatial separation of the respective rivers provides people with a sense of escape from everyday experiences, from roles and rules which are institutionalized in work settings.

The coupling of the three Southbanks with their city centers is constantly being improved by the construction of new pedestrian bridges, new landmarks, more active and welcoming frontages and clearer sightlines, yet in the cases of Brisbane and Melbourne, connections back away from the riverfront and through to other neighborhoods have been made significantly worse. Behind the riverfront, the boundary is sharply defined. These precincts have become, to varying degrees, colonized as tourist enclaves of the city center (Judd and Fainstein 1999).

The precinct of Southbank Brisbane, as an example, 'is disconnected in a completely uninteresting way from . . . South Brisbane behind it; it is planned in thin slivers of water/path/water . . . and [has a] maniacal directionality to the city' (Macarthur 1999: 180–1). This disconnection consists of several layers of impressive defensive fortifications: first and most ironically the dull blank backsides of four massive cultural institutions and their loading docks; second by wide roadways, an elevated railway and dedicated busway, their stations and thousands of parking spaces, and third by the 400m-long bulk of a Convention Centre. Despite recent dramatic improvements to one key pedestrian corridor, most pedestrian links through to South Brisbane, the city's most vital cultural and artistic hub, are awkward, invisible or absent. Although Southbanks prosper in their marketed image as ideal pedestrian environments, the changes which have enhanced connectivity to them at the metropolitan scale have also reduced it locally, marginalizing these settings as a whole, making it difficult for people to wander, explore and mix. Walled residential enclaves, internalized shopping malls, and office towers which are vacant on evenings and weekends also threaten accessibility to urban waterfronts (City of Melbourne 1997). In many cases new buildings also overshadow the enhanced public promenades which are intended to be the main goal of the exercise.

Although each of the three Southbanks has many programmed leisure facilities, the playful acts described here do not conform to the scope of leisure which is so carefully designed into these kinds of escapist zones (Hannigan 1998). The players eschew passive reception of these safe, predictable, commodified spectacular landscapes, often literally turning their backs on them. Southbank is clearly not 'edgy' enough, and people find ways to transcend its limits. These unexpected play activities are more creative, active, self-directed and self-sufficient. They are thus sometimes difficult and risky, and they reduce the potential profits of vendors. These forms of play nonetheless rely on the carnivalesque atmosphere which has been created in these settings. The simulation of a playful 'place apart', where rules are relaxed, ends up stimulating desires it cannot contain, leading to unbridled forms of play.

These play activities also involve close sensual engagement with specific site conditions, particularly water. In waterfront leisure planning, water itself is, symptomatically, all too often framed reductively as something to view, and it can be difficult to reach or cross river embankments. While marginality may be about getting away from the majority and getting off the beaten track, it is also about getting close to novel and sometimes dangerous sensations. These activities on the margin highlight that play is not only about looking at something.

Unexpected, disorderly activities on these Southbanks reveal disobedient marginality (*paidia*) within structured marginality (*ludus*), answering Lefebvre's call for a lived critique of modern capitalist society's segregation of times and spaces for leisure 'which are as clearly demarcated as factories in the world of work' (Lefebvre 1991b: 227). One way to interpret these dialectical outcomes is to accept such manicured spectacular landscapes as they are, noting that heterotopic moments remain possible because people will always bring new desires and creativity. Campo (2002) for example notes people finding ways over, under and through barrier fences which cut them off from Brooklyn's deindustrialized waterfront in order to pursue informal leisure activities there.

Another approach is to take from these examples inspiration for the future planning of cities. If the escapism of crossing boundaries and taking risks is a popular and important way of playing, then perhaps it is inappropriate and undesirable for marginal zones like Southbanks to themselves have rigid physical limits or controls on behavior. In particular, perhaps waterfronts and other leisure precincts need to be more thoughtfully interlinked with other inner-city activity areas, functionally, temporally and spatially, as the Situationists intended by their concept 'unitary urbanism':

> Unitary urbanism acknowledges no boundaries: it aims to form a unitary spatial milieu in which separations such as work/leisure or public/private will finally be dissolved.
>
> (Debord 1996b)

Testing the edge

Play as a form of risky bodily engagement with physical edges in space is typified by the vertiginous play of cyclists, skateboarders and in-line skaters. The edges where cyclists and skaters play include both vertical surfaces which can be pushed against, and raised edges where movements on wheels involve sudden, thrilling exposure to shifts in height and speed.

On London's South Bank, trick cyclists have the opportunity to play within several different dimensions of marginality (Figure 6.20): a leisure zone separated from the city, the edge of its pedestrian space, and the concrete edge of an object. In an interesting contemporary inversion of the Situationists' entreaty for citizens to rip up the (functional) pavement to discover the (playful) beach underneath, here the cyclists have discreetly removed the layer of rubber padding which made the edge of this planting bed comfortable for all-weather sitting, exposing a flat metal plate which is ideal for them to slide along. The edge is now less comfortable but more fun. Play at this site is a highly visible performance because it is adjacent to the South Bank pedestrian promenade where it makes several bends right next to the popular Tate Modern gallery.

Skaters often play against the edges of Melbourne's Town Hall Plaza, which is backed by a flight of large steps and faces onto busy Swanston Walk. Two in-line skaters get up speed across the plaza, jump up and grind the base of their skates along the edge of the first step and down the bumpy edge of the adjoining sculpture (Figure 6.21). Skateboard grinding involves pushing the board against an edge such as a step. Momentum holds the skater up there. The small surface contact minimizes friction and allows them to move quickly and smoothly along the edge. The skater's attention becomes focused entirely on the detailed qualities of the edge; he or she can feel the boundary of the space 'press back' against them (Borden 1998).

Melbourne's city council later installs raised metal lugs at regular intervals along this ledge in an effort to deter skateboarders from playing here. Play in urban public space often has a dialectical context: players try to change the built environment to meet their desires, but managers of public spaces also react by making changes which suit some kinds of leisure and inhibit others. But in this instance, the skateboarders do not go away. Skaters do

Figure 6.20 Cyclists practice tricks on a concrete planter edge which has been stripped of its seating pad outside the Tate Modern gallery, South Bank, London.

not skate because it is easy. They enjoy taking risks, testing their skills and developing new moves. The variety of physical challenges in the landscape actually stimulate their desire, heightening the fun and engendering creativity. Three teenagers jump their skateboards onto the lower metal face of the sculpture, which is at an 80-degree angle to the ground (Figure 6.21). Vertical scuff marks all over this base indicate this is a popular practice. The same counter-play measures have proliferated throughout most public areas in inner Melbourne, and many cities worldwide, with similar effect. In front of the State Library, skaters now weave between the lugs, or jump their boards up the stairs, or risk injury by jumping off landings and flipping their boards in mid-air. On Southbank, skateboarders even prefer one section of landscaping where the walls, ledges and benches have anti-skating lugs over a mirror-identical section which is free of them. At the Tianjin Garden, there are metal lugs set into every concrete step and ledge which could conceivably be reached. Yet skateboarders have reached inconceivable heights: they gain speed down the adjacent footpath and leap onto ledges more than 600mm above the pavement. Chipping, scratching and paint marks on these 'skate-proofed' ledges and the dramatic bending of one lug, cast from 10mm thick steel, from repeated heavy impact illustrate skaters' willingness to transgress these restrictions either by deft control or by brute force. The bending of

Figure 6.21 Skating along smooth and rough edges: Town Hall Plaza, Melbourne.

this lug exemplifies Lefebvre's argument that play is a 'functional' activity which 'produces' space:

> it is in the nature of energy that it be expended – and expended productively, even when the 'production' involved is merely that of play or of gratuitous violence. The release of energy always gives rise to an effect, to damage, to a change in reality. It modifies space or generates a new space.
>
> (Lefebvre 1991b: 177)

Such traces bear witness to new functions for edges, advertising the feat to others. Skateboarders also leap for slipperier and more risky edges within the Tianjin Garden, including an inclined handrail. Through this playful act, an element which is intended to make the use of public space safe and functional becomes an opportunity to experience the risks of rapid movement and intense bodily negotiation of the edge (Borden 2001a). Rather than moving slowly down steps, skaters jump off them or up them, or slide quickly and smoothly across them.

Further challenges are presented by the long main staircase in front of Melbourne's State Library. This is a complex geography of inclines, drops, barriers and smooth ledges with many options for engagement. Late one night, there are many in-line skaters on the library's upper terrace. Novices do slow circles while experienced skaters glide down the inclined surface, accelerating under the tug of gravity, riding out the difficult terrain at high speed. Their smooth motion loosens up the physical striation of the steps. One skater is skilled enough to glide forward or backward down the main steps, and even to jump backward from the top terrace, land half-way down the first flight, continue backward, swivel at the landing and then roll forward down the second flight (Figure 6.22). This skater is going around to others who have watched or commented on his trick, asking whether they can go down backwards, goading them. One novice rolls over to the edge of the terrace and looks down, but decides she cannot adequately manage this difficult and dangerous feat. Another man, trying to learn through trial and error, exclaims to others around him: 'I've got to keep falling until I get this!' There are many high-speed accidents, but those who fall generally smile as they lie sprawled on the pavement, communicating to others that they are still having fun.

Playful use of edges of more utilitarian settings can be just as extreme. Outside an anonymous Collins Street office building, the arc of wheel marks above the steps and extensive scraping on the foreground handrail trace a trajectory where skaters leap into the air from the upper terrace and manage to meet the handrail and then slide down its incline (Figure 6.23). Wheel scratches leading off the meter-high end of the terrace highlight that this is another popular spot to jump. Along the length of this terrace, the footpath gradually falls away with the topography. Skateboarders roll down the footpath and along onto the terrace and then turn to jump off at whatever height gives them the desired level of thrill and risk. The freedom to choose allows them to gradually push their limits. Skate marks starting and ending at different points along this test area reveal different levels of speed, strength, control and confidence among users.

The smooth, level, 'safe' open space commonly found at the top of public staircases is important because it allows skaters to prepare, to accelerate and to orient themselves for takeoff. Play arises from sudden engagement with the drop, but people want to be able to manage this exposure, to time it, to choose their angle of attack. The man who leaps backward down the State Library's main steps must skate in from one side to generate speed and then do a deft half-turn just before jumping off. The office terrace in Figure 6.23 is shallow, and the skater's approach must be parallel to the edge rather than perpendicular; there is the extra challenge of arcing at the right moment to cut across the threshold through the air between the planters and handrails. The depth of flat, open space at the bottom of staircases is also important. Skaters generally fly from thresholds where there is somewhere clear to land, particularly because they do not always land safely. This area is a place to enjoy the momentum that comes from the descent, to decelerate and to be on display.

Skaters tend to choose marginal locations which are not in use by pedestrians, including dead-ends which do not lead anywhere and ledges which are too high for pedestrians to walk up onto. Skaters tend to frequent steps of office buildings in the city's business precinct outside of the normal rush and lunch hours. In the case of the library, which is often busy, there are several flight of steps, each much wider than is necessary for pedestrian volumes. The excess space can be used luxuriously for play, and for other people to sit and observe. The importance of an audience in stimulating skaters' displays of daring should not be discounted. This is not to suggest there is never friction between skaters and pedestrians. Skaters emerging quickly and unexpectedly from the edge of a space are often on a collision course with people walking past (Figure 6.24). They have little time to avoid each other.

Steps, ledges and handrails can be found around most buildings and plazas in inner cities. Skateboarders' choices of specific sites reflect the design and variety of the edges which are available. One important factor is choice: for vertigo to be playful demands controlled encounters with the uncontrolled. Skaters want to regulate the risks they are exposed to.

Nonetheless, risk remains the key ingredient that makes all these forms of engagement with edges a desirable and absorbing escape from normal bodily experience. Risk exists because skaters relinquish the slow and steady connection to the ground which is characteristic of pedestrians. They either balance precariously on the edge, or project themselves free of it. They escape the preconceived bodily trajectories around which the urban landscape is designed, and discover other potentials which are latent in its edges.

Figure 6.22 'I've got to keep falling until I get this!': skating backwards down steps at night outside the State Library, Melbourne.

Figure 6.23 'A little more risk than can easily be handled': skating surfaces of office building steps, Collins Street.

Skating is an unintended consequence of these sophisticated landscape designs, with their many edges and their many carefully conceived functional elements. Playful uses of the physical terrain such as skating challenge the social conventions which spatial boundaries imply. Skating can be read as a lived critique of overdetermined fantasy landscapes such as Southbank with their rigid framing of appropriate, largely passive forms of leisure activity (Hannigan 1998), but also as critique of a functional approach to architectonics (Borden 2001b). In both senses, skaters' playful encounters foreground active engagement with spatial conditions. They highlight two general ways in which physical boundaries provide potential for the thrill of vertigo. First, to explore a space's physical limits is to make the most of its potential for action. At the boundary, space can be experienced up close, through the sense of touch, as material and not just as visual form (Borden 1998). The sense of touch is not passive; it has a reciprocity. Skaters' actions provide opportunities to sense their own body and test their bodily capacities up against spatial boundaries. Skaters' slipping, sliding, rolling and grinding negotiates the substance and texture of edges and boundaries. Second, skaters' leaps from edges highlight that vertigo involves transgressive actions which exceed limits. Beyond a boundary lies the start of some other kind of space, more difficult and dangerous. Skaters' leaps take them out

Figure 6.24 Leaping from an edge fitted with anti-skating lugs: Casino Promenade.

into open air or down a slope. Observations of other kinds of playful activities show that risking the edge of water can also be exciting.

Not many people test the edges of public spaces in the same ways that skaters do. Yet the retroactive attempt by urban designers to control and constrain their playful behavior is at odds with the very idea of marginal, escapist, transgressive spaces set apart from the working city, and also at odds with citizens' own aspirations for play.

In the first sections of this chapter the focus was on cases where people chose to locate themselves at the edge of public space because it is a site of passivity, greater individual control and lower risk from which to either observe the activities of others or discreetly play. The condition of marginality indicates that boundaries can enhance people's freedom to behave differently. In other instances, people choose the boundary as a site for display because boundaries can heighten visibility. The physical properties of boundaries and their level of permeability – to bodies, to visibility, to sound – vary greatly. Rather than being constraints on playful relations and behavior, such variations are opportunities which help to stimulate and support a very wide range of play. Boundaries are elements which can be mobilized to enhance and structure playful behavior under different social contexts.

People make the most of boundaries by strategically positioning themselves behind them or in front of them, by pushing them, adding to them or moving back and forth across them. Users of public spaces assess the roles that other strangers are playing in urban encounters partly through their perception of how these strangers position themselves within settings. Distance is a very important cue to social interaction among strangers. Distances and levels of engagement can be confirmed or adjusted by boundary markers such as paving patterns, street furniture, areas which are sunlit or shaded or otherwise enclosed, or changes in level. Boundaries such as railings, windows, high steps, streets or rivers permit some kinds of engagement but can restrict movements, physical contact or spoken communication. Design and management that hold people back at the boundary become part of the context for play, lending it distinctive flavor. Sometimes the atmosphere of play arouses desires which lead people to ignore social convention and transgress boundaries.

People also play by engaging directly with boundaries, often as a way to intensify their bodily experience: sitting on boundaries, obscuring them, transforming them or breaking them down. Tangible, static boundaries of urban space such as walls and steps provide something to play against. Tangible but loose boundaries such as the water's edge also provide risks which people engage with playfully. In all cases, outcomes of action vary according to the relative strength of boundaries and of people's resolve to violate them. Boundary maintenance is a constant game.

Chapter 7

Thresholds

A threshold is a point on the boundary between inside and outside that can be opened. A wide range of perceptions, movements and social encounters become possible there. As Norberg-Schulz notes, 'the opening is the element that makes the place come alive, because the basis of any life is interaction' (1971: 25). A threshold is also a restricted space; its design always constrains people's behavior and their perceptions (Hillier and Hanson 1984). Many different architectural elements distinguish inside from outside and mediate people's passage between them: doorways, turnstiles, colonnades, marquees, porches, terraces and stairways. Although these elements are designed to serve practical and ceremonial functions, thresholds present distinctive perceptual, behavioral, social and symbolic affordances which also give rise to a great variety of play.

The observations which follow illustrate five distinct ways that thresholds can mediate experience, starting from the most straightforward mechanical physics, and leading on to conditions which are more socially contingent, complex and imprecise. The threshold is a constrained site which gathers people together, channeling their movement, focusing their attention and forcing them into close contact with others. It is a passage, a transitional place where people spend time. It is a space set apart from the wider public realm where people can control their level of exposure to others. The passage across a threshold also frames people's emergence from private spaces into the public realm; it is a site of new stimulations. And yet a threshold always remains both-and, in-between, inside and outside, a loose mix of two different environments.

Convergence

Four thresholds in central Melbourne where a great number of playful events were observed were the entries of major public buildings: Flinders Street Railway Station, the General Post Office, the State Library and Parliament House. Each has a large formal facade fronted by a wide staircase, with a grand arch or colonnade marking the entry. It is hardly surprising that much playful activity occurs in front of such buildings. A great number and variety of people use these buildings. Even though their entries are often small and crowded, the generously scaled forecourts which they open onto make them likely settings for informal use.

In front of the wide staircase of Melbourne's central Flinders Street Railway Station, skateboarders roll back and forth, sometimes trying to jump up the steps. Even though the luxurious width of the stairs allows skaters to avoid conflict with practical users, friction does exist. One skateboarder races across unexpectedly in front of oncoming pedestrians and grinds along a step's edge. Two teenagers kick around a Hacky Sack right in the path of people entering the station (Figure 7.1). They are watched by a small group who have been approaching the station from the pedestrian crossing. The group's physical orientation toward the contest has precipitated their interest in it. Inside the station vestibule, a Maori does a brief informal version of the *haka* war chant and dance in front of the turnstiles, symbolically blocking the path of commuters leaving the city (Figure 7.2). A threshold can frame unplanned encounters which stimulate play because it puts other people's superfluous activities between someone and their goal, potentially distracting them from practical action.

Figure 7.1 Threshold as urban theatre: view from Flinders Street Railway Station steps.

Figure 7.2 Maori man dances a *haka* (war dance) at the turnstile entry to Melbourne's main railway station, Flinders Street.

London's Leicester Square is fronted by the thresholds of numerous entertainment venues: three major cinemas which screen world premieres, a nightclub, and the city's half-price theatre ticket booth. Nearby lie the Shaftsbury Avenue theatre district, the National Gallery and the Royal Opera House. These permanent spectacles attempt to capture tourists' imaginations for commercial gain, but on the way in and out of such venues, tourists have to pass through an extraordinarily busy, distracting pedestrian zone. The entertaining ambience spills over the thresholds of buildings and into the public square itself. The entertainment spectacle which attracts the crowds gives rise to derivative forms of performance, which ultimately help nourish it. Street performers dress up and stand on pedestals, pretending to be statues of film characters (Figure 7.3). This performance on the threshold draws upon three distinctive features of the location. First, the memorial statuary so common in English parks: examples here include local sons Shakespeare,

Figure 7.3 Street performers pretending to be statues, Leicester Square, London.

Hogarth, Newton and Charlie Chaplin. Second, this street theatre is a consequence of the strong local demand for professional actors. Third, they feed on the fantasies of filmgoers. An audience, on its way toward the formal venues and keen to consume fanciful performances, encounters these figures, as characters (almost) stepping down from the screen. Onlookers can walk up to and around these actors and even touch them. This public setting enhances the tension and the immediacy of their performance. In the context of close encounters with a diverse, unpredictable, excitable audience, staying completely still is perhaps the most difficult act of all.

Figure 7.4
Street theatre
'The Bells' preceding
performance in
indoor theatre
(right), Marlene-
Dietrich-Platz, Berlin.

One Saturday evening in Berlin, a touring troupe performs a piece entitled 'The Bells' to hundreds gathered in front of the Stella-Musicaltheater on Marlene-Dietrich-Platz, near Potsdamer Platz. The bellrope-tugging actors lunge out into their audience, leaning perilously close to the spectators. Behind the performers, above the building's entry, is a massive billboard for the current show: Disney's 'The Hunchback of Notre Dame' (Figure 7.4). As the act finishes, program sellers emerge from the Musicaltheater, its doors just opening for the evening show. The representational link is purely accidental. However, the slope of the plaza toward the doorway and the

building's awning roof are not. The temporal link from street theatre to indoor theatre amplifies the intentional spatial connection.

These events highlight the distinctive spatiality of social relations at thresholds. They are natural gathering points, bottlenecks where many people's paths must converge (Lynch 1960). In such settings, chance and risk are always present (Goffman 1982). Thresholds are likely places to meet both friends and strangers. What is distinctive about social encounters on major thresholds is their frequency and intensity. Particularly when strangers have to negotiate doorways and queues, they are unexpectedly forced into close proximity, increasing bodily exposure and thus tension. Spatial convergence also offers the potential for more dramatic or confrontational encounters. Thresholds are scaled and designed to serve flows of pedestrians. Playful practices such as the war dancer, singers and skateboarders interrupt this instrumental function, pushing into people's line of vision and movement. Although people are forced to converge on the steps of Flinders Street Railway Station, close encounters there are generally playful and enjoyable because people retain control: there is enough space to avoid any involvement, to go around, and still reach one's goal.

People also often gather at thresholds willingly. Urban leisure settings are typically the scenes of dense flows of people who are already outside their serious work role. People heading into London's and Berlin's theatres appear to enjoy being intercepted by free, informal, more physically engaging performances which take place outside their doors. These events extend people's escapist, liminal experience in time and space, beyond the controlled conditions of the venue. In the case of 'The Bells', there is a symbolic link between the performances on the threshold and the stage as well as a functional one.

The thresholds of major urban railway stations serve as thresholds at the scale of the city as a whole (Lynch 1960). One's arrival in the metropolis is a dramatic rite of passage. Large numbers of strangers who are engaged in quite unrelated activities find themselves gathered together here according to the rhythms of the trains. Exposure to new experiences and heightened sensations is likely, and this can stimulate playful opportunities:

> Railway stations are characteristic places for dense and varied as well as anonymous and fleeting encounters, in other words, for the type of interactions which were to mark the atmosphere of life in big cities, described by Benjamin as overflowing with excitement.
>
> (Habermas 1997: 229)

Regardless of the physical tightness of such places in terms of bodily movement, the sheer number of people and phenomena they contain establishes a threshold condition in the sense of testing people's tolerance for new perceptions (Simmel 1997).

Another dimension of the liminality of thresholds becomes apparent when zooming out from single building entries to the scale of the urban plaza and urban block. A city street which provides views and access onto a wide variety of activities is most stimulating to the senses and generates the richest urban life (Gehl and City of Melbourne 1994). For urban space to provide a concentration of the city's symbols and experiences requires constraining the width of individual thresholds. When the thresholds of many buildings with different uses are pressed close together in the street, new experiences and unplanned juxtapositions of social groups and behaviors are more likely (Jacobs 1961). In Leicester Square and Marlene-Dietrich-Platz, it is the variety and collective intensity of the theatres, cinemas and nightclubs which make these squares so lively, drawing a great number and mix of people at different times. Urban areas with many narrow, active thresholds are most conducive to people aimlessly strolling around and 'window shopping', which can lead to strangers meeting each other unintentionally and having non-instrumental interactions.

Street blocks in Melbourne's CAD which have a greater number of separate frontages are also generally the sites of more numerous play activities. The redevelopment of Melbourne's Southbank was based on maintaining a continuous, high-quality 14m-wide public waterfront promenade for its full length (City of Melbourne 1997). Problems relate to the land use pattern fronting onto this pathway, particularly later phases of development further downstream to the west. The Crown Casino complex is highly internalized, with few entries, and lacks any public control over the mix of uses, while the Melbourne Exhibition Centre is a monofunctional, access-controlled space. Each building occupies 500m of frontage (City of Melbourne 1997; Sandercock and Dovey 2002). As with all urban streets, the most fundamental threat to pedestrian vitality on the waterfront is the permitted maximum frontage given over to a single use and sole management (Gehl and City of Melbourne 1994) – essentially the monopolization of spatial experience within what is supposed to be a public realm. Eckstut (1986) argues for the need to design waterfronts on the same principles as inner-urban streets, as coherent, mixed-use pedestrian environments. As Jacobs (1961) noted, small-scale blocks and a fine-grained mixing of primary and secondary uses are both necessary to maintain vitality; flows of people and pools of uses mutually underpin each other.

The Passage of Time

Most people use doorways and stairs only to get from one place to another. Thresholds are often planned for only fleeting occupation, yet people often spend time on them. Some instances of play at thresholds highlight the temporal dimension of behavior, as people occupy and modulate this brief time in transition by playing. It is interesting to consider the reasons why they linger, the ways they extend their visit, and the distinctive behaviors associated with moving across a threshold.

After a morning tea-break outside a Melbourne office building, a worker heading back inside with colleagues accidentally drops his balled-up paper bag. He slides forward to grab it and feigns a basketball shot before stepping through the revolving door. This play is a last attempt to use up excess energy and live more intensely, before returning to the physical passivity and mental discipline of desk work. Alexander et al. (1977) suggest that the design of transitional space at building entrances is important for helping people to lose their 'street behavior' – their tension and their blasé detachment from strangers – so they can relax and 'settle down' into a secluded, private environment (Alexander et al. 1977). They note that 'it is the physical changes which [create] the psychological transition in your mind', and that it is therefore desirable to mark this transition 'with a change of light, a change of sound, a change of direction, a change of surface, a change of level, perhaps by gateways which make a change of enclosure, and above all with a change of view' (Alexander et al. 1977: 552). Playful behavior highlights the role of such design elements as steps, doorways, awnings and courtyards, as well as the auditory and visual porosity of buildings' facades, in effecting psychological shifts as people move across a threshold between the exuberance of public play and the relative passivity of private spaces.

The transition from a private realm back out into public gives rise to a great variety of playful possibilities. A large group in their twenties stand around talking under a cinema marquee after a movie. One man grabs a woman around the waist and picks her up squealing and laughing. Hundreds of people emerge after Good Friday mass at a Catholic church; most stand around talking in the church's large fenced courtyard. Some are playing, releasing pent-up energies after a prolonged period of bodily and emotional restraint. There is some fake sparring between friends. One teenager puts his hands on a friend's shoulders from behind and temporarily boosts himself up. Five teenage girls try to run along as a group with their arms around each other's shoulders; they get about 10m before there are too many obstacles.

In church or during a film, simulative play is experienced through a relatively passive, highly structured consumption of sounds and images. Attention is directed and physical interaction is minimal. People may emerge with their bodies restless and their perceptual faculties in a highly stimulated state. Both conditions make play at thresholds a possibility. When participants eventually take their leave to go home, their paths will diverge. It is only in the times and spaces immediately surrounding these events that everyone in the group can interact freely. What eventuates depends on people's mood, social circumstances and the physical setting. If the proportions and facilities of a threshold are generous, it provides for a comfortable pause in the movement across it: somewhere to shelter and to reflect before heading out.

Time spent crossing thresholds is often prolonged by events which occur there, even when these events present no physical obstacle. The Europacenter shopping complex sits on one corner of Berlin's Breitscheidplatz. One Sunday evening a rock band performs outside the building, next to its main entry doors. Hundreds of people watch, either standing or sitting at adjacent café tables. The performance is very polished and energetic. Some people dance enthusiastically. Their liberation from the passivity of watching reflects one of the band's selections: 'I want to break free'. Similarly at Alexanderplatz, Berlin's historic center, Christian evangelist youth groups often sing and dance in front of the Galeria Kaufhof department store. One sunny Saturday there are thousands in the plaza, and a group plays African drums near the doorway. Two couples exit the Kaufhof together; one man does a few little rumba steps to the beat. In these examples, someone occupies a strategic position between pedestrians and their destination, encouraging the passers-by to extend the time they spend at the threshold, presenting an opportunity to escape their everyday responsibilities.

A crowd of thousands gathers at Alexanderplatz to view a major solar eclipse in the summer of 1999. Some people are compelled to keep working during the eclipse, though even for them the threshold can become a site of escape. A youth working in a fast-food restaurant facing onto Alexanderplatz steps out of its front door, whips the special glasses out of his pocket and takes a quick look at the eclipse. A co-worker rushes out after him, grinning, and also has a quick look. The first youth conscientiously looks around to retrieve used food trays: he finds a legitimate pretext to come outside and momentarily join the throng in their admiration of the natural spectacle. He manages to ever-so-slightly bend the rules of his employment, creating a free space-time, using instrumental demands on his time and location as leverage for an escape from instrumentality.

A great variety of playful behavior can be observed outside leisure-oriented facilities such as cinemas, theatres, nightclubs, casinos, cafés and churches. When entering, crowds must converge and wait for a programmed event. Afterwards, although people can quickly disperse, they often linger outside in groups. There is a close relationship between highly structured, instrumen-talized consumption indoors and unregulated, liminal activities immediately outdoors. After a prolonged period of mental and bodily discipline and passivity, people are keen to release pent-up energies. Spontaneous, active forms of play often arise: games of tag, running, jumping on each other and similar kinds of horseplay.

A space apart

Another range of play activities occur at thresholds where people can relax in a setting which is comfortable and expansive. Some thresholds are specifically designed to make the passage between inside and outside gradual and leisurely, sometimes including several intermediary spaces such as semi-private foyers, doorways, wide landings and generous steps, as well as forecourts which frame views of grand entrances. In the case of large buildings with wide frontages, reduced pedestrian flows at certain times of day provide a surfeit of public space which is then open to appropriation for various uses. In these kinds of settings, people have space, as well as time, for a gradual transition, for lingering and for non-instrumental social interaction.

At Melbourne's Flinders Street Railway Station, people often sit to rest or to meet friends on the stairs outside the main entry. This includes many who have not traveled by train. Those gathered here watch playful behavior and street performers on the open paved area in front of the steps (Figure 7.1). These flat open areas allow performers freedom of movement and allow their audience to relax, comfortably immobilized on the steps. People are already moving more slowly when they walk down steps, and if they see something interesting happening below, they often stop to watch. Informal performances also occur on the steps themselves. University students going through an initiation ritual form a choir spread up the steps to sing 'Jingle Bells' (Figure 7.5). Sometimes strangers passing by or sitting and watching become active participants in these playful performances.

In the doorway on an inner-Melbourne laneway three office workers take a cigarette break. This doorway has a thick stone frame, with a deep landing at the top of five steps. A woman standing against one side of the entry talks animatedly to two colleagues opposite. Meanwhile a man passes by on the footpath below, heading away from her. After he passes, the woman comments to her colleagues, then steps down to the footpath and mimics his walk. Continuing the parody, she looks fixedly ahead, serious; then coming abreast of the doorway, she suddenly swivels her head to view where she herself had been standing and puts an exaggerated beaming smile on her face. All three laugh. One colleague re-enacts his own version. The threshold space frames the interaction among these three workers, but the public realm is the necessary ingredient which triangulates, providing novelty which arouses playful action (Whyte 1980). The doorway shapes her encounter with the stranger. This threshold provides a 'back' region 'where the performer can reliably expect that no member of the audience will intrude' as well as a 'front' region 'where the performance is presented' (Goffman 1959: 98). As the man moves along the street, the constrained views both in and out of the doorway, which protect the back region, suddenly expand, heightening the spatial and temporal compression of the two individuals' contact and increasing its dramatic tension. The passer-by

Figure 7.5 Informal choir singing on steps of Flinders Street Railway Station, Melbourne.

unexpectedly notices the woman as he passes the building entry, and reacts spontaneously to this close engagement. Stepping out of his normal role as a plodding, detached pedestrian, his overstated smile is a small impromptu performance for her.

These examples illustrate the use of thresholds as a 'space apart', physically distinguished from both inside and outside, available for play (Huizinga 1970). These thresholds also serve as special places for play because they lie between two very different kinds of social space. The examples show that movement across thresholds involves gradations of perception, regulation and exposure between fully private and fully public. Thresholds are complex spaces which structure a great range of social and bodily relations. Indoors there is relatively tight control over behavior and visibility. The smokers have come to the threshold to escape the regulation of their use of time and space. The exuberance of the woman's actions expresses this freedom. At the threshold inhabitants are also first able to perceive random strangers and be perceived and engaged by them. Busy thresholds in the city are sites of frequent encounters. At thresholds, an observer is 'likely to experience a sudden rush of information – a sudden dilation of his view and exposure too – which may (or may not) suit his intentions' (Benedikt 1979: 58). The woman smoking in the doorway is suddenly noticed by the passer-by when he comes abreast of the building entry, much like the sudden encounters which often occur at street intersections.

Thresholds are sites of tension because exposure means the risk of losing control. The three smokers are able to enjoy their encounter with the stranger passing by because they can control their own level of exposure. Their playful mocking of him is possible because they can manage their separation from the public realm. The woman smoker is close to the footpath, but elevated one meter and in shadow. Standing against one reveal of the doorway, she has a close view of people passing by. She is unthreatened by the glances of those coming toward her because of her separation within a distinct, defensible space, and she can intently examine those heading in the other direction. Through small movements of her body, she uses the threshold's depth to finely adjust her level of stimulation and risk from public exposure against the control and security offered by the private realm. She and her co-worker choose to come down from their defensible space and imitate the stranger after he has left, when the street is empty, so they are not themselves exposed to either ridicule or reproach.

Staircases and landings outside thresholds provide a space apart, define a place to wait and rest which is separated horizontally and vertically from both the regulated indoors and the constant movement of the street (Huizinga 1970). The brownstone stoop has long been an important setting for informal socializing and play (Dargan and Zeitlin 1990). The tiering of steps can offer a large audience a good overview of all the action in public space. Visibility works both ways: steps can become stages or seating. Sitting is a relaxed mode of spending time in public space, a position in which the body is at rest but attentive to what is going on around it. A raised threshold optimizes the publicness of playful performances there, because it enhances exposure. It increases the number of people who have a clear line of sight from the street below, even when the setting is very congested, and allows the performers to be seen from a greater distance. Stairs thus frame relations between audiences and actors which add to the tension of the latter's performances (Caillois 1961; Whyte 1988). Steps structure social distances between strangers and calibrate their exploratory encounters: each tread is an increasingly engaged threshold. When people choose where to sit on flights of stairs in public, they are able to adjust how close they are to public scrutiny, challenge and unpredictability. Because steps are designed to facilitate movement between spaces, they also provide for an easy transition between passer-by, seated audience member and player. Thresholds allow people to regulate their exposure to the unfamiliar and to risk in a number of ways, to manage the problem of overstimulation. People have varying attitudes and desires in relation to the freedom of public space. Not all individuals want to take themselves to the limit.

Extended, repetitive, 'unproductive' uses of thresholds challenge the minimal, instrumental conception of urban space (Gilloch 1996). The activities observed show great variation in the temporality of people's experiences of thresholds. Because they are points of necessary convergence, thresholds are places where people often have to pause, reflect and change direction. For these very functional reasons, people can find themselves unexpectedly distracted or delayed at thresholds. Playful events which spill across thresholds show that people do not always minimize their contact with others. Musicians outside Berlin's shopping centers stimulate people's obvious interest in escaping the practicalities of their everyday tasks. They are not blocking the entry, but stand in close proximity to the portal, hoping people in the flow will be distracted and will extend their necessary activity of shopping into an optional one (Gehl 1987), as when the man's stride takes on a rumba beat. In Jacobs' (1961) terms, playful, disruptive acts are secondary, derivative 'uses' which rely on the primary use inside a building's doorway for their vitality.

The act of passage

Another crucial reason for the varying duration of the uses of building thresholds is that thresholds form an interface between two quite different spatial, perceptual and social realms. Outside in public space people are suddenly exposed to new and diverse stimuli, to unstructured encounters with strangers, to freedom, anonymity and risk. Inside in relatively private spaces, ambience is regulated and social behavior and encounters are more carefully structured (Norberg-Schulz 1971; Markus 1993; Dovey 1999). The office worker playing basketball with his rubbish underscores the sharp opposition between work and play. His behavior marks the end of a special time on the threshold. He recognizes the threshold as a point where a space and time apart from instrumentality ends. People play at thresholds because it is their first and last chance to act upon the freedoms and inspirations which urban public space provides, where they have the opportunity to 'be themselves'. At the moment people cross thresholds which mediate between private and public realms or between indoor and outdoor space, they often make the most of the experiences which are possible there. Their movements can be exuberant, expressive of escape from spaces where they have defined roles and commit their attention to specific tasks, of freedom from controls on their use of time and space. The physical and temporal constrictions on behavior at thresholds further heighten their complexity, but despite constraints, thresholds are sites where conventions get loosened through people's diverse playful behavior.

When playful social behavior involving movement across a building threshold is recurrent, it can become ritualized and symbolically charged. The ongoing production of the representational potential of thresholds can be illustrated by the ritual of formal wedding photographs taken in central Melbourne, which very often capture the betrothed couple crossing the thresholds of various city buildings. Many connections between marriages and threshold spaces have already been identified by Shields (1991). He examines how the physical characteristics of Niagara Falls, a popular honeymoon site, provide a metaphorical expression of the social distinctions framing newly-wed couples, representing their liminal identity and their private and public selves. Liminality, taken from the Latin word for threshold, is an anthropological term for the intermediate, playful stage in 'rites of passage', the progression from one social status to another (van Gennep 1960; Turner 1982). Wedding photographs in Melbourne similarly use thresholds as a symbolic landscape which expresses this liminality. The threshold is the scene of a dramatic change in people's status, both experientially and symbolically.

Wedding photographs framed in doorways emphasize the liminal, transformative nature of the wedding ritual. Standing on the threshold implies that the man and woman are poised between the roles of single person and couple. Bourdieu (1990: 281–82) argues the threshold is 'the site of a meeting of contraries'; it is 'the point where two different worlds [the public world of men and the domestic world of women] meet in order to "fertilize" each other' (Stavrides 2001: 2). The couple emerge from the private realm, with its connotations of sexuality, in order to present themselves as a respectable couple in the public gaze. This action is symbolic of the new couple's respectability. The doorway frames a dialectical transition between the personal and the social.

Many wedding photos are taken in the doorways of old stone buildings, which define the partners as stepping out of the past. This underscores the evolutionary nature of the marriage ritual. The thresholds of old buildings combine a sense of a liminal moment with a sense of history. The timelessness of the setting is a reminder that the extraordinary event of the wedding serves to reproduce the strength and permanence of social structure. In addition, the thresholds of secure institutions such as Parliament (Figure 7.6) and banks (Figure 7.7) lend social legitimacy to this ritual, just like a church does. The use of old, expensive-looking buildings as backdrops also expresses the luxurious value of the wedding. Doorways and lobbies of a building often have the most ornate detailing. The expansiveness of some threshold spaces, particularly Parliament House with its massive colonnade and staircase, also expresses luxury, lending a sense of grandeur and prominence to the wedding.

Framing photos within doorways, porches, at the tops of steps, and most exuberantly on a window ledge (Figure 7.8) emphasizes that the couple are on public display but still set apart, held up above the public – a status suggested by the German word for wedding, *Hochzeit* or 'high time'. Figure 7.8 shows that the liminal state of the couple allows them special dispensations. Their costumes indicate to the public that they are performing a special role. Hence they do not have to behave according to their usual

Figure 7.6 Getting married is a big step: wedding photos outside Parliament House, Melbourne.

Figure 7.7 Wedding as historic event: historic bank building, Collins Street.

Figure 7.8 Creating new metaphors: wedding photos on window ledge of Melbourne Town Hall.

station in life or the expected uses of spaces. The wedding party's posing in the window is a ritual inversion of the proper, practical, decorous use of the space. The general public stands by, tolerating and even celebrating their moment. Shields (1991: 156) points to the predominant image of Niagara Falls as 'a site of the carnivalesque, a landscape of kitsch and popular parodies of dominant aesthetic and moral judgment'. He suggests that its natural and built features lend it a permanent air of social transgression and inversion. Posing on the window ledge of the Melbourne Town Hall suggests that the participants see this urban threshold in a similar light. This spatial practice defies easy interpretation. It is representational, however it cannot readily be reduced to words (Lefebvre 1991b). This example illustrates the power people have to create new social discourse through their actions within the built environment. This performance extends the meaning of both wedding and window.

Many wedding photos in front of buildings show the couple on their way somewhere. The threshold is a place of movement, and flights of steps outside doorways dramatize this sense of flow, leading the eye across the picture

(Figure 7.6). Sometimes the groom stands one or two steps ahead of the bride, suggesting he is leading her into the public sphere. Steps emphasize that the couple are in a directional movement, symbolizing a social progression. Many social conditions differ between the inside and outside of a threshold; wedding photos at thresholds are framed at physical thresholds to emphasize that things are likewise different after the wedding. It is 'a place where there's just no turning back', as the couple's new status becomes sealed (Shields 1991: 144). Long flights of stairs cascade ahead of the couple; getting married is indeed a big step. Newly-weds are generally pictured heading away from buildings and down steps, down into the everyday, public world of the city streets. One example shows the couple in the bridal car parked in front of a doorway. They are ready to be transported on the journey of their life together. Things start to move faster once they pass across this threshold.

The popular use of the grand staircase fronting Melbourne's Parliament House, which on weekends is often the site of simultaneous wedding shoots, frames couples on the precipice of a huge incline, exposed to physical vertigo and to the excitement and licentiousness of the city which stretches out below. A comparison can be drawn with Shields' (1991) observations on newly-weds' visits to Niagara Falls. Both are liminal initiation rites where the couple's exposure on the threshold of a sublime, intoxicating, erotic setting serves as a test of their passions. In both settings, couples face the temptation of 'a ludic explosion: a high of repressed sexual energy' (Shields 1991: 153): confronting the intense allure of these sites serves to sanctify the love of the wedding couple, by demonstrating that their love is stronger than the greatest distraction.

Photographs are sometimes taken where a narrow lane meets a wider street. These lanes are small scale, with old buildings, brick or stone paving and little or no vehicular traffic. Some of the lanes used provide service access to warehouses; they are not luxurious or distinguished. Nevertheless, the backdrop of an old cityscape brings a sense of history to an important moment of personal history. Two other values are represented through these backdrops: a sense of intimacy (through small scale) and a symbolization of being in public space. The couple are generally posed emerging from a constrained, intimate space into wider, brighter public space which offers a range of different directions and views. This is a threshold at a larger scale, framing the same dialectic between privacy and publicity.

The taking of wedding photographs, often thought of as a highly ceremonial social event, actually helps to loosen up the meanings of spaces. Through their wedding photographs, people seek out and perform a variety of social meanings which lie latent in the spatial properties of thresholds: intimacy and publicity; progression and irreversibility; transgression, security and permanence; distinction and grandeur. By gathering such meanings together within the picture frame, wedding photographs help constitute the liminality of the wedding. Through their playful actions, wedding couples are not just acknowledging, but actually producing social meanings, and inscribing them within built form. The theatricality of wedding parties' performances suggests that the photographs are not necessarily a reflection of identities whose nature is already well understood, but rather an important means of discovering new selves, through encounters with space and with the urban public.

Many of the observations elsewhere in this chapter demonstrate that even in the absence of ritualized understandings, people's diverse, playful acts of passage across thresholds can create temporary conditions of intensity, transformation, escape from convention, the elevation of status and the blurring of social categories and rules. These acts help to create liminal moments in everyday life and allow people to explore and develop their identities.

Blurred space in-between

Play around some thresholds is diverse because the physical and social conditions prevailing at them are so varied. Some thresholds are clearly demarcated and controlled, but in other cases play arises precisely because the threshold is quite nebulous and ill-defined.

On Melbourne's main shopping street, Bourke Street Mall, retailers project intensive messages out across their thresholds, to force them onto the awareness of passers-by. People's exposure to any single business is brief, partly because of narrow shop frontages, and there is much competition for attention; there is little discreet about retail thresholds in the city. One retailer's logo dances around on the footpath itself, projected by a laser where pedestrians frequently look down to ensure their safe footing. Loud pop music spills from the open front of a music store, saturating public space and distracting people's attention away from their conscious objectives. This music attempts to awaken desires in the body so as to stimulate consumption (Crawford 1992). The same compilation of songs cycles endlessly, highlighting that most people's exposure to this music is incidental and fleeting. The purpose of the music is to pique an interest, not to satisfy it. But commercial interests can only stimulate desire, they cannot dictate what actions arise from it. Some passers-by seem to ignore the sound entirely, refusing to let it focus their attention. Others sit around and receive it passively.

An athletic man in his fifties dances energetically just outside the store, for fun (Figure 7.9). This man's act expresses people's freedom in responding to the sensory stimuli which are compressed together in urban space. Sound is immaterial and ephemeral; its pleasure lies not in being possessed, but in being experienced. Here the body's exposure to sound arouses a playful exploration of the body's capacities. When the beat of the music is slow, the man slowly writhes his body. As the music speeds up, so do his steps,

Figure 7.9 Stimulus spills over the threshold: man dancing in front of music store, Bourke Street Mall.

and his moves get more exuberant: 'Thriller'-style shoulder jerks, attempts at moonwalking, disco spins. His vertigo is clearly inspired by the atmosphere that spills out across this threshold. Sometimes he accompanies the tune on a harmonica. His responses suggests that music 'found' in the public realm is not a sacred, finished piece of work, but remains open to interpretation. The man escapes everyday, practical behavior by attuning his body to the music's rhythms. He is within the audible threshold of this store, but by remaining in public space, he remains free to respond to the music however he wishes. He generally has his back to the store. Rather than yielding his attention to the merchandise which the store is trying to promote, the man makes use of the sound to draw attention to himself.

The man's enjoyment is enhanced and given shape by the presence of many strangers in the public space of the Mall, who witness his act, comment upon it, and take part in it. Pedestrians do not seem to mind passing close by the man; some talk to him. He invites passers-by to join him, and several do. His enthusiasm rubs off, and this relies on the man being poised on the threshold of a public space which many others are using for their own purposes. Later the dancer increases his exposure by moving further out into the Mall, opposite a pedestrian lane which leads into it. Many who pass close by the man cope with this friction in space use through 'civil inattention' (Goffman 1980): they give the man a smile and then have a discreet word to each other once they have passed. He calls out to a couple who have been watching him from a distance, and later goes over to chat with them. They praise his dancing effusively. The woman wants to join in, but her boyfriend is too self-conscious to take part in such a display. The boyfriend points out that three young boys, who the man has been trying to cajole, have finally started dancing. Two of them take turns making funny dancing moves, laughing. They mimic the man for each other's amusement, and keep looking across at him. When he heads back over, they stop. Their freedom to play depends on not being closely watched or approached.

Like the smokers who move in and out of the doorway, the dancing man draws upon stimuli both inside and outside the threshold. Moving further out into the public realm expands the threshold as a field of play, increasing the prospects for the man to engage in dynamic interactions, rather than just putting on a display which is viewed passively from a distance.

The video game arcade is another private leisure environment which extends beyond its threshold and contributes to the atmosphere of play in public. Pedestrians passing one Melbourne arcade at lunchtime stop to watch two players dance on sensor pads, following a disco beat which keeps accelerating. The video screens show little: the competitors' attention is on the music, the onlookers' attention is on the competitors. The youths finish with a synchronized jump. The large audience applauds, then suddenly dissolves and moves on. This wide, open facade on a busy pedestrian route provides good views of the action, temporarily distracting passers-by (Figure 7.10). This video arcade is well located on the west side of Swanston Walk, which has most of that street's commercial activity, and hence most of its pedestrians. Only one short block north of Flinders Street Railway Station, it is one of the first sights of interest for those entering the city. It sits on the north corner of Flinders Lane, and the glazed side wall facing this lane gives the oncoming pedestrians a long preview of the video game action.

Although the video game exists in a private world apart, it extends its challenge across the threshold into the public realm. The newest games are always strategically placed just inside the open threshold. Their large screens are turned either toward the street or parallel to the footpath, aiming to attract not only players but also spectators. The attention of people who are hurrying past, focused on instrumental tasks, is drawn from the street into the open arcade and further into the imaginary realm of the screen, which is a stage for competition. The siting of the game machine at the threshold puts both the player and field of play on public display. For the players, video games pose a competitive test of the limits of coordination, often involving simulations of risky, bodily active sports such as shooting, skiing and horseracing. Video games are a form of leisure which is stigmatized by (and for) adults as a waste of time, money and effort. The barrier

Figure 7.10 Distraction across an open threshold: video game arcade, Swanston Walk.

between watching and playing is highly permeable. Energetic teenage players provide vicarious pleasure to adult pedestrians (Figure 7.11). For onlookers, the game is a simulation, a fanciful escape. These thresholds offer glimpses of exotic life possibilities beyond the public realm. The urban setting adds the tension of performing in front of a diverse and spontaneous audience, spurring the players' efforts. The arcade threshold exposes and stimulates playful possibilities without providing much of a regulatory function: watching crowds spill both into and out of the private space.

A third example highlights the playful possibility of a threshold which is permeable for both music and people's movement. A teenage couple dance enthusiastically outside the doorway of a small café in a narrow pedestrian lane. The front facade has open, unglazed windows, and loud music spills from within. The couple dance briefly, then bashfully hurry back to rejoin friends sitting and watching just inside. The form of this threshold space makes it feasible for the couple, when inspired by the music or encouraged by their friends, to quickly move to a space where energetic dancing within earshot of that music is possible, and just as quickly to return to the interior and relax in relative privacy. Additionally, the audience does not have to make a special effort to watch the action outside.

Figure 7.11 The liminal threshold of a transgressive space: video game arcade, Bourke Street.

In these examples, open facades frame public performances which are stimulated by sound and motion generated inside private spaces. The man dancing oriented himself to the music store's threshold to receive and mobilize specific kinds of sensory and social stimuli in his play: music from within a private realm and exposure to a public audience. The threshold of a video arcade exposes and stimulates playful possibilities without providing much of a regulatory function; it offers the possibility of distraction. Such thresholds are liminal in the sense suggested by Zukin (1991): there is a seamless, frictionless possibility of entry into the pleasure zone, a blurring which allows people to 'forget' that a social boundary exists (Figure 7.11). The provision of the escapist world of the computer screen, the total absorption of the players and the safe performance of their extreme bodily actions are all facilitated by a controlled environment. The arcade is a world apart which prevents serious realities impinging on adventurous play, whereas the openness of the arcade's threshold illustrates the dependence of the games on spontaneous desire to participate or watch, and the thrill of public exposure. People often seek the tension of such sites which are 'defined and yet not too defined' (Alexander et al. 1977: 349). The significance of threshold permeability is underlined by the case of the couple dancing outside the café. A longer duration of exposure in the public space would heighten the risk of an encounter with passing strangers. If the boundary between resting, choosing to play and watching was not so porous, this activity may not have taken place. Goffman (1959) notes the importance of this immediacy in his description of the threshold between a back region and a front region:

> By having the front and back regions adjacent in this way [connected 'by a partition and guarded passageway'], a performer out in front can receive backstage assistance while the performance is in progress and can interrupt his performance momentarily for brief periods of relaxation.
>
> (Goffman 1959: 98)

People utilize the thresholds between social realms to control the rhythm of their social interactions. This control enhances their freedom to choose to play.

The design of the video arcade's threshold frames peripheral views which can stimulate playful behavior. Glass walls provide a preview of what will be seen, heard and felt up close at the entrance. The game screens fall within the normal cone of vision of even those passing pedestrians whose gaze remains fixed directly ahead. The passer-by's response is similar to the 'Gruen transfer' in shopping malls: 'the moment when a "destination buyer", with a specific purchase in mind, is transformed into an impulse shopper, a crucial point immediately visible in the shift from a determined stride to an erratic and meandering gait' (Crawford 1992: 14). In shopping malls, adjacent attractions, such as handbag shops near shoe shops, are never a matter of chance; the shopping experience is carefully scripted (Crawford 1992). Yet on urban streets with many thresholds, with relatively unregulated juxtapositions of uses and styles, one can never be sure what will come into view through the next display window or doorway.

These observations illustrate how the social liminality of thresholds can arise from a softening of distinctions between inside and outside which is made possible by wide, transparent and open frontages, floor surfaces continuous with the footpath, advertising and merchandise placed in the street, and awnings and verandas which extend the interior out over the footpath. Publicly accessible thresholds of stores do not neatly separate public from quasi-public settings. Such open thresholds do not filter sensations according to the practical needs of passers-by. Views, music, smells, breezes and actions from both realms extend across the threshold and overlap. Passers-by become aware of the escapist atmosphere spilling out from these thresholds. The flashing lights and music at the video arcade, music store and café are highly stimulating because of their intensity and variability. The dancers provide movement and actively encourage others to participate. Whether such sensations are distractions or attractions depends on the changeable attitudes of individuals; they merely offer options to respond.

The negotiation of thresholds

Thresholds at all scales are places of movement, but that movement is not as straightforward as might be imagined. Observations show that people's uses of thresholds are tremendously varied and not always efficient or practical. The flows of people across thresholds vary in their rate and direction; there are also significant differences in the ease with which people move and the risks involved; thresholds frame careful, calculated moves as well as unexpected ones. In their diverse playful movements across thresholds, people engage with their physical, social and representational parameters.

Some thresholds are sites for very public forms of play because large flows of people have to pass through a constrained space. Thresholds channel people's movement, focus their attention and intensify sensation. In many cases they force people into close contact with strangers. Goal-directed convergences of commuters, shoppers or audiences present opportunities for playful, slightly confrontational displays by others. Steps at thresholds enhance the potential for seeing and being seen. Playful encounters in doorways demonstrate that physical constraint can serve as the stimulus for many novel and enjoyable experiences. Examining movements in the opposite direction, thresholds are also points of emergence from a controlled indoor environment. Space opens out, providing new stimulations and new possibilities for action. Play is a response to the euphoria of this release. Whether a threshold is experienced as a convergence or divergence, it is a site which brings the physical and social world into focus, heightening awareness of opportunities.

Release from spatial and social constraints indoors also means an increase in exposure, in the possibility of unplanned social encounters. People at play on thresholds make small-scale movements in relation to various architectural features there, including windows, columns, steps and shadow, in order to subtly adjust their perceptions and calibrate their own public exposure on a variety of registers: visual, auditory and bodily. Inevitably, people slow down or stop as they move across building thresholds, particularly when a door or a change in level has to be negotiated. Slowing down increases the possibility of being distracted by views, objects, other users or nearby performances. Focusing on the transitional nature of threshold use reveals that people can compress or stretch this liminal experience in time. They adjust the rhythm of their movement across thresholds, to control when and for how long they step outside the everyday and the expected. People thus use thresholds to bring a measure of structure to their encounters with otherness. Such control is essential if such interactions are to be playful.

But threshold spaces are always, by their very nature, only partly defensible and also partly unregulated and disordered, shared with strangers, other activities and unfamiliar experiences which are always in motion. Thresholds can be both physically and existentially slippery. Much of the liminal, playful potency of thresholds lies in the dramatic tension of a continually shifting balance between exposure to 'open' space and personal control and in unexpected juxtapositions of actions and meanings. In the liminal space of the threshold, the distinctions which society writes onto space sometimes become blurred and dissolved. Thresholds 'separate and connect at the same time' (Stavrides 2001: 1). There is ambiguity about how people should behave at a threshold and where their attention is focused. Thresholds offer both opportunity and risk largely because of their blurred, indeterminate in-betweenness. They sometimes precipitate playful experiences precisely because they expose people to 'a little more risk than can easily be handled' (Goffman 1972: 63). Risk makes people focus their efforts and attention and makes their play compelling. The frequency with which people linger at thresholds suggests that people savor the unpredictability and the overlap of roles and sensations which are available there.

Both the liminal qualities and the general representational potentials of building thresholds have a documentary celebration in wedding photographs, where married couples act them out. Like weddings, thresholds are special settings for the play of meaning because they are both-and, between social categories. Thresholds support a 'culture of negotiation' (Stavrides 2001) where identity can be reconfigured because they frame liminal spatial conditions of transformation, intensity, contrast, escape and risk. Wedding photographs also show how people's behavior lends new meanings back to thresholds. The sanctity of the meanings of marriage gives sanction to unusual actions. Though not what thresholds were planned for, liminal, playful activities test the boundaries of what is acceptable and what is desirable about behavior in threshold spaces, loosening up the rules and expanding the common understanding of the value of these sites.

People cannot always occupy thresholds on their own terms; this complex social and physical geography often requires hesitation and social negotiation. The diversity of uses of urban thresholds is not without difficulties or conflicts. Spontaneous, exploratory, impractical and dangerous uses of thresholds can be a bodily threat to people's instrumental need for passage. Security officers often chase away those who play and ask those sitting to move aside and keep the threshold clear. Yet because building thresholds are generally designed only for momentary, fleeting occupation, and because they must provide access directly from a public space, they will always remain open to temporary appropriation by various members of the public for new uses.

Chapter 8

Props

Paths, intersections, boundaries and thresholds are all reasonably large-scale properties of the urban spatial structure within which people act. These elements shape how and where people move and encounter each other within public space. There is also a microgeography of built elements that structure human experience and movement within the body's reach and that the body can move around. Such elements may easily be overlooked as a part of the environmental structure because they are small and because people tend to perceive them as being within space rather than shaping space. Public artworks, play equipment and street furniture are three kinds of fixed objects placed within public settings to make them more comfortable, contributing to their aesthetics and function. Yet these objects also make possible and stimulate a variety of non-instrumental, exploratory and risky forms of play behavior.

The term 'prop' emphasizes that such objects sit on the stage of urban public space and are made use of in a variety of social acts, in particular playful performances for the benefit of strangers. These props frame spatial relations between people, such as triangulation, and lend meanings to people's performances. Some props are intentionally representational and provide a catalyst for people's simulative play. Whether deliberate or not, all props are challenging physical landscapes which invite close exploration with the body. Engagements with these elements of urban form are very often tactile. Some props are consciously designed to stimulate physical play. Even in such cases, solo players and groups creatively explore and expand the range of types of play which play equipment makes possible.

Public art

One way of playing with props is to engage with the meanings which they represent. The public artwork *The Three Businessmen Who Brought their Own Lunch: Batman, Swanston and Hoddle* (Figure 8.1) consists of three very thin, life-size bronze statues wearing business suits and carrying briefcases. They stand on the kerb at a street corner and stare quizzically out at the city. People invent many different ways of playing with these life-size statues, all of which pretend they are real people. They stand arm-in-arm with the figures, hug them, imitate their stiff stances and their comical facial expressions. They shake their hands, pick their noses and pat them on the belly. Many playful engagements with these statues are transgressive of behavioral norms, including hitting, kissing and impolite touching. They are performances of imagined social relations with strangers which are inappropriate given the anonymity of urban society, particularly considering these are businessmen.

People explore the sculptures' physiques through both vision and touch, and discover what the statues can 'do'. Their playful contributions imagine new roles and identities for the statues. The fine details of the statues' hands and faces provide much to develop play around, and support lasting physical interventions. A balloon is left attached to a hand. In winter one figure is given a woolen hat. All three of the figures have been designed with mouths pursed into deep circular holes. A woman passing by puts her lit cigarette in the mouth of the rear statue, and she and a friend laugh, recognizing that passing strangers are confronted by her contribution (Figure 8.2).

The woman's intervention provides humorous comment upon the specific social and spatial situation of the figures. *The Three Businessmen* are waiting anxiously at a tram stop; it is only natural that they should have a quick cigarette. They are inanimate, but they are still able to smoke. The act of adding the cigarette proffers a link between smoking and idleness: smoking is something people do when they are doing nothing, when they are killing time. Smoking and giving cigarettes to statues both relieve boredom. Putting cigarettes in the mouths of statues is wasteful, pointless, bad. Public statues usually expresses society's higher ideals; adding a cigarette transforms this sculpture into a promotion of something profane. While statues, like social ideals, are meant to endure, the added cigarette also introduces an ephemerality which undermine the permanence of moral pronouncements through sculpture.

Many of the tensions of urban life are written into *The Three Businessmen*. They appear tense, harried, expectant. Their formal dress and posture contrasts with the humor of their exaggerated features. The looks of surprise and apprehension on their faces, their frail bodies and unsteady, tilted stance suggest an inadequacy. They are figures of fun to be approached and interacted with; their meanings are not intellectually threatening or distancing. Their identity is completed and extended through interactions with real people. These sculptures provide an opportunity for people to make their mark on the lives of an imaginary other. It frames a proactive role within a close encounter, in the undemanding mode of let's-pretend. The mouth holes in particular offer an easy opportunity for people to transform meaning.

Figure 8.1 *The Three Businessmen Who Brought their Own Lunch*: intersection of Swanston Walk and Bourke Street Mall.

Figure 8.2 Profaning public art: *The Three Businessmen Who Brought their Own Lunch*.

These statues can easily be drawn into simulative play because they are carefully scaled within an urban theatre. They are located on the footpath at eye level and in easy reach. They stand at Melbourne's busiest pedestrian intersection, on the southwest corner of the Bourke Street Mall and Swanston Walk, where many people stop to look around, to cross the street or to wait for trams. The statues stand exposed in a widely spaced row. Their appearance emphasizes a reciprocity between prop and actor, making people's approaches less confrontational. People are free to inspect them up close, to touch them, and to treat them irreverently. Bodily engagement with them is reasonably effortless, while the high pedestrian traffic ensures the results are noticed. Many tourists pass here, so there is always someone nearby for whom the statues are a novelty. The orientation of the statues, facing the wider public space, optimizes the visibility of performances using them.

In some cases, play involving the statues happens without an audience. This suggests that people engage in simulative play not just as a display to others, but to test their own bodily skills, as an escape into fantasy, and even just for its own sake, for the pleasure of the bodily experience. Even if only for a brief minute, the figures totally absorb people's attention, diverting them from their instrumental responsibilities and concerns. This artwork makes possible escape from everyday behavior because, like most public art, it lacks 'function' in the strict sense; it does not help achieve any specific practical outcome. The message this sculpture is supposed to convey is also unclear.

A similar representational public artwork in Berlin is a black metal statue of a man sitting on a park bench in one corner of Breitscheidplatz. A black man sits next to him and polishes him. He has a container out to collect donations for his 'impersonation' of the statue. The common sight of performers everywhere who pose as statues has produced particular expectations about what makes simulation entertaining. The amusing result here is that bystanders often believe the statue is a real black man. Even when his untidy 'cousin' has left, people stand and stare to see if he will move. They are afraid to invade his personal space. They drop money in the dish to reward his posing. But not all people are fooled all of the time. The deceptiveness of a simulation can itself provide the basis for play. On another bench in a different corner of the same plaza is a sculpture of two elderly women talking. Tourists pose for photos next to them. A young man walking past feigns punching one of these statues in the head, then nurses his hand, pretending to have hurt it. These two examples show that planned displays can engender unforeseen responses. Some people pretend they are unaware of a simulation; others show their appreciation of a simulation which is actually a 'real' statue; some try to make money out of the confusion. What people think about these statues is constantly reshaped by how other people are acting toward them.

Another sculpture in human form which prompts play is a hollow metal mould on Melbourne's Southbank Promenade shaped like the rear profile of a small, naked figure (Figure 8.3). Approximately 30m further back from the river edge, the front half of the mould shows the figure is a woman. A man and woman examine the rear half of the mould together, and the woman briefly puts herself inside it. An hour later another couple explore it, and again the woman climbs inside. Play actions around this prop engage with its representational aspects, but they also take pleasure in tactile experience. The women 'become' the statue by filling its absence; they assume its pose and its point of view. The statue provides an opportunity for the women to reshape their identity, to momentarily escape from themselves by simulating someone else. The other playful dimension of this behavior is the close bodily encounter with the sculpture, feeling one's way into its landscape. This statue frames the vertiginous thrill of a tight space. The challenge is to endure constriction of the space all around the body.

Another example of vertiginous play occurs around a family of statues on the footpath just outside an office building who seem to be peering through the window into the foyer (Figure 8.4). A father walks past with four young children, who notice the statues and run over to them. Rather than mimicking the statues or responding to their poses, the children squeeze themselves through the narrow gap between the statues and the glass, edging their way through it. The children's squeals of delight indicate it is the close bodily negotiation of this compressed, dark, mysterious space which provides a thrill. The window side of the gap is smooth and difficult to locate, whereas the statue side is hard, uneven and potentially painful. This act of play centers on immediate, tactile experience, and has no regard at all for the representational characteristics of the artwork.

A different kind of representational prop which often gets used for play is the sculpture *Architectural Fragment*. This piece, sunk into the footpath of Swanston Walk in front of the State Library, represents the upper corner of a neoclassical building, with a portion of the word 'library' inscribed on its frieze. But *Fragment* does not get played upon because it is a Library; indeed, people never seem to play with the representational qualities of this object. The sculpture's symbolic elements face toward the street, and away from pedestrians. Due to this orientation, the symbolism lends itself primarily to distant, fleeting, passive contemplation, as suggested in the city urban design unit's own assessment of *Architectural Fragment*:

> Is it sinking in cultural decline or bursting through the pavement in renaissance? [The artist's] half-seen bluestone fragment, replicating a corner of the library's classical facade, can be interpreted in a number of ways. What is not in doubt is that this eye-catching work is both an amusing and thought-provoking addition to the streetscape.
>
> (City of Melbourne 1997: 15)

Figure 8.3 Tactile interior: body mould, Southbank Promenade.

Figure 8.4 Prop frames a narrow space: facade of office building, Lonsdale Street.

The city's assessment ignores the possibility that the sculpture could catch more than the eye and provoke more than thought; that it could become a prop in an act of physical play. The 'top' of the sculpture is a smooth inclined plane continuous with the footpath. Skaters and cyclists frequently practice their skill at rolling up and banking around this slope (Figure 8.5).

Architectural Fragment functions as a prop because it sits on a stage where an audience is gathered together to watch actors perform. The inclined top faces toward a heavy flow of pedestrians as well as people looking down from the grassy slope of the library forecourt. This adds to the thrill and interest of playing here. The detail of this face of *Fragment* simulates the material and paving pattern of the footpath. This other representational content of the sculpture subtly encourages the vertiginous form of play captured in Figure 8.5 by suggesting that people could try to use the sculpture as if it were a footpath. Part of what enables cyclists to move beyond contemplation of *Fragment* is its adjacency to function. The normal, practical footpath in front of the prop is essential for the build-up of speed to launch oneself onto the face of the sculpture.

Architectural Fragment operates on different levels to serve different needs. Some meanings stimulate direct playful engagement; others serve only

Figure 8.5 Vertiginous footpath: *Architectural Fragment*, outside the State Library, Melbourne.

contemplation. While the primary connotation of this prop as 'library' is obvious, it is not clear what people should do with the cornice of a building. The sculpture is both representationally and physically impenetrable. This representation of 'cultural decline' also provides an incline that allows cyclists to reach new heights of vertigo. The sculpture is two-faced, and its playful use by cyclists and skaters illuminates the tension between two different readings.

Play involving these various public artworks illustrates the multiplicity of opportunities which people can find in them. Artworks which have human form appear to stimulate a great breadth of imagination from passers-by. This is partly a matter of scale and physical complexity, and partly relates to people's awareness of the human form and their readiness to act in relation to other bodies. The intended and most obvious representational meanings of props do not delimit the range of possible play actions which involve them. This is unsurprising, considering that play, as an escape from instrumentality and convention, pursues new, unexpected and transgressive experiences. New meanings for these physical forms are discovered and acted out through close bodily engagement.

People also use the concrete physical attributes of props to nourish their play. They discover and activate the potential for vertigo which can be latent within simulative objects. These forms of play treat a prop as just another part of the urban landscape. Playful uses of *Architectural Fragment* highlighted a discord between playful representations and playful functions: the meanings of the urban landscape always depend on the desires of the user.

It is only when such public artworks are 'on stage' that there are likely to be playful performances which communicate new meanings for them. *The Three Businessmen*, for example, is sited on a busy public thoroughfare which provides a flow of potential actors and exposes their acts to a large audience. By contrast, the clustered family of statues outside the office building, the body mould on Southbank and the fragment outside the library all literally turn their backs on nearby pedestrian flows. It is hardly surprising that their symbolic potential is often overlooked. The use of *Architectural Fragment* by cyclists illustrates that space for approach, the angle of the sculpture to the ground and places to sit and watch are important to displays of skill.

Planned play equipment

A fountain adjacent to the riverfront promenade outside Melbourne's casino (Figure 8.6) is an example of a public artwork specifically designed to

Figure 8.6 Playing with chance: casino fountain, Southbank.

provoke playful bodily engagement. Similar installations exist in many popular tourist sites throughout the world. This water sculpture is composed of numerous circles of jets set flush into the pavement; the jets are sometimes idle, but most of the time they squirt upward in unpredictable sequences and to varying heights.

Passers-by often gather around to passively observe the spectacle of the dancing jets. The fountain stimulates various senses through its constantly changing motion. The water rises and falls in salvos, sequences, alternations, pauses. Visual stimulation is enhanced at night by multicolored lights. The fountain's hydraulic action also produces an exciting array of sounds, from the abrupt whoosh of the jets, the hushed ripple of the swaying columns of water when they hang suspended, the hissing spatter when they are caught in a breeze, and the exuberant pregnant splash when they collapse onto the pavement. The fountain's unpredictability, its physical design and the expectant audiences that gather here also stimulate varieties of play which are more closely and actively engaged with the moving water. Its carnivalesque energy spills over and encourages spontaneous, unprogrammed, risky behavior. The fountain is 'on stage', adjacent to a busy pedestrian promenade and surrounded by seating. Members of the public often step out onto this stage and begin unscripted interactions with the moving water and with each other. People's many and various engagements with this fountain include all four basic types of playful experience, and range from very primitive experiences of *paidia* to highly complex and reflexive social games of *ludus*.

Moving very close to the fountain to get a more immediate experience of its power, unpredictability and danger is a form of vertigo. Two young boys who have been told they are not allowed to get wet move as close as possible to the edge of the line of the jets, and stand immobile as water erupts 6m high right in front of them and falls again. One of them manages to catch some of the water in his hand as it falls. They experience perceptual vertigo. Another boy, already soaking wet, timidly places a baseball cap over an idle jet. When it fires, the hat jumps 2m high. The design of the fountain eliminates the physical infrastructure, so that people can have an unmediated engagement with the water itself, jumping, falling, collapsing, breaking. The water appears as a solid object which rises with force and then suddenly softens and vanishes. The fluidity of the fountain's material lends a special thrill to the experience of touching this sculpture. The continuity of the pavement through the fountain encourages close encounter by minimizing the effort. There is a permanently dry area inside each of the fountain's three rings of jets, and people delight in stepping 'inside' one of these rings, where they can remain reasonably dry while being completely surrounded by a raucous upsurge of water, momentarily trapped in a world apart. The randomness of the water's dance and the unpredictability of the wind ensure this experience retains an element of chance.

The columns of water which spring from the fountain's jets are, like *The Three Businessmen* statues, at ground level, right in front of people. They are also scaled relative to the human body, being particularly well fitted to actions of the hand. People can stand outside the fountain and grasp it, or the water can rise up around them and hit them. The fountain has a sense of interiority: one can be inside this prop and hence be highly involved with it, wrapped up in it. There is a certain playful reciprocity about people's interactions with it: they dance with the water as if it is another performer.

People enjoy the challenge of trying to pass through the length of the fountain without getting wet, either by outrunning the shooting jets or understanding their pattern and dodging them. Whether through mental concentration or purely physical effort, this is play as competition against chance. A 3-year-old boy stands close to where the water jets splash. He is transfixed, squealing with delight. He joins his father, who has carefully walked into the center circle (Figure 8.6). They step around within the jet area, paying careful attention to where the jet spurts. Their play reveals a growing familiarity with the pattern underlying the apparent tumult of the water. In some cases, people attempt to beat the fountain by imitating it. Five children try to run, following the sequence of the jets, in a three-looped figure-eight. This play is not purely tactical: it is also pleasurable because it has its own special rhythm in time and space, offering an escape from rational patterns of movement. In other cases of simulation, people use the dancing shape of the fountain as the basis for escapism, constructing a fantasy that they are somewhere else or someone else. Two girls dance a hornpipe over the jets, stepping in a sequence that avoids the random spurts. Their absorption in the timing of the dance, and the idea that they are dancers, perhaps even sailors on a slippery deck, brings added tension to the game of avoiding the jets of water. A boy standing in the middle of the fountain pretends he is the sorcerer's apprentice from the film *Fantasia*, conducting the arcs of jets to leap ever-higher. Such play does not concern itself with the risk of getting wet. All his concentration is applied to a focused flight of fancy.

The aforementioned kinds of play all focus primarily on the characteristics of the prop itself, and on the individual's relation to it through vertigo, competition and simulation. Other kinds of play use the fountain to give shape to playful relations among people. People experience vicarious pleasure watching others play in the fountain. The sensory delights of people running and laughing and the risks they are taking add to the stimulation of watching the fountain itself. Easy visibility into the fountain for audiences who stay at a safe distance is crucial to this relatively passive form of interaction. In one instance, parents look on anxiously as a boy of 4 wanders across the fountain, oblivious to the danger, but miraculously escapes getting soaked. He often lingers over spouts, which the parents start to find very funny. They have resigned themselves to watching a game of complete chance, and ultimately enjoy escape from their own everyday concerns thanks to their child's own nonchalant tempting of fate. For the child, moving into the wet zone serves as a way of escaping parental control: they cannot do anything about his behavior without sharing the same risk that he does. After he makes one pass successfully through the fountain, his parents scold him, but he goes again. He defies authority, confronts the risk, and is vindicated. He makes rude gestures at them from the center, and they laugh heartily. Another family with teenage children stand discussing the fountain as it shoots skyward in solid walls of water. The daughter dares her father to run across, and the other relatives chime in. The father stands a meter from the nearest jet, yells 'Now!' and mocks running into it, but does not. He taunts them, raising their expectations by pretending. The actors in both these examples consciously play with the tension felt by an audience. The fountain becomes a testing ground for individuals doing things they should not do or things they would not normally do. They play with the fountain as a way of testing the social expectations of their relatives, through a variety of dares, feints and transgressions. The prop is deployed in a competition between people. Tension and interest is maintained because the risks of this fountain are hard to calibrate.

Sometimes strangers also encounter each other in informal games framed among the fountain's jumping jets. On New Year's Eve, a hot summer night in Melbourne, thousands of people mill around on the casino promenade. Hundreds form a dense circle around the vividly lit fountain where several teenage girls are playing, soaking wet. A boy successfully runs through the fountain from end to end without getting wet. When he tries a second time, two of the girls block his path shoulder to shoulder. He has to make a detour around them and gets hit by one water jet. The girls add their own competitive skills to the arbitrary, chance obstacle presented by the fountain itself, respatializing the challenge and redefining the strategies necessary to win. Some youths in the crowd clap and cheer the boy. He runs again and the girls attempt to block him. He charges between them, and they grab hold of him. Next, four older teenage boys try to run through. The girls catch two of them, and everyone gets wet (Figure 8.7). The large crowd follows the contest attentively. A 5-year-old boy stands in the middle of the first circle of jets. It seems he is intending to chance the girls. They watch him and flex in anticipation; he loses his nerve.

This modified game of 'British Bulldog' in the fountain is a superlative example of public play. It happens at a special, liminal time, in a place set apart from seriousness. It involves competition, chance and vertigo; it inspires people in the audience to step forward in simulation of those who have gone before; and it engenders encounters among strangers. It is a game that emerges spontaneously. Although it is derived from other known games, it has a loose structure, and continues to evolve. Many attributes of the context inspire this exploratory social game: the spontaneity and danger offered by the random fluctuations of the fountain, the different zones of space which

Figure 8.7 Game of 'British Bulldog' on New Year's Eve: casino fountain, Southbank.

are defined by the water jets, the stimulation of the colorful lights and sounds; the euphoric sensation of being soaking wet; the gathering of potential players with unknown desires and capacities; the expectant hum of a huge crowd of onlookers. Some onlookers are sufficiently inspired that they became participants. The motion of the water continually reframes the relations between actors, onlookers and risk, generating dramatic tension. The 'usefulness' of this prop for play continues to develop through the kinds of playful actions that people invent for it.

A majority of play within the fountain is created by children and teenagers. This public artwork stimulates forms of children's play which harmonize with the desires of adults who may come to the casino to gamble. The fountain's seemingly random sequences of spurts forge a conceptual link between chance, sensory excitation, playful exuberance and the experience of release. Although the fountain has an abstract, changing shape, it is nonetheless representational: it functions as a metaphor for the games and machines inside the casino. The carefully orchestrated risks of engagement with the fountain provide an illustration to potential gamblers that taking chances can be both fun and safe. Its high-profile public setting emphasizes that risks and games are also fun for passive onlookers. The fountain supplements the casino's interior leisure attractions, making this a landscape that presents risks the whole family can enjoy.

Street furniture

Public artworks are designed and sited to distract the attention of passers-by and to stimulate and support play within carefully circumscribed parameters of risk. Street furniture, by contrast, is not generally designed with play in mind; it is designed to be safe and efficient. There are, however, tensions between designers' conceptions of function and everyday social practice. Playful forms of movement and body poses and playful social contexts expand upon the possible uses of street furniture. People who wish to play often put these pragmatic objects in the service of vertigo and competition. Because they are not designed for play, the tension, risks and thrills of play on street furniture are all heightened.

There are a large number of fixed metal bench seats along Swanston Walk, the Bourke Street Mall and other footpaths throughout the Melbourne CAD. One night in front of the State Library, a cyclist manages to ride up the seat and the curve of the backrest of such a bench and up onto the very top of it. He balances precariously astride the two back-to-back benchheads, poised high above the ground. He pauses, gazes out down the long grassy slope to the lights of the city beyond, and bounces a few times on his suspension. Then in one continuous motion, he rolls backward down onto the seat, lifts the front wheel high in a 'wheelie', and pivots around 90 degrees. He then expertly 'bunny hops' the bike on its back wheel back down to the ground. The cyclist uses his wheels to feel his way around the bench, making the most of its potential, exploring its miniature landscape. The continuous curvature of this bench's profile, with a bullnose at both the low edge and the crest, and the slight play in the wooden slats lend themselves to his sinuous movement, to smooth acceleration and deceleration. A group of young men outside a video game arcade on Bourke Street play quite differently on a bench: they stand on top of one and wrestle with each other, playing a variant of the game 'king of the mountain'. The bench here frames a physical competition between the teenagers.

Another item of street furniture put to use in a range of play activities is bicycle racks which have the form of tubular metal arches fixed into the footpath. Two teenagers climb on such an arch to see over a large crowd watching a theatre performance in a store's display window. As a third teenager reaches up, trying to join his friend on top of the arch, the friend keeps lifting his arm out of the way so that he has nothing to hold onto. Winning this game is too easy, and soon the friend relents and helps him up. Competition is transformed into cooperation as they attempt to both balance on the narrow, smooth curve to have a view of the performance (Figure 8.8). The pole against which all three teenagers steady themselves

Figure 8.8 Arch as prop in a playful balancing act: Bourke Street Mall.

can barely be reached; there is a tension between the desire to focus on the show and the need to exert effort just to stay up. Their attention to this instrumental need generates its own interest and excitement. Nearby, a stranger simulates these teens' strategy by climbing up on a bollard.

During Berlin's Love Parade, the size and density of the one-million-strong crowd makes viewing the Parade into an act which requires tremendous effort, creativity and risk. People climb high on every available utility box, signpost and lamppost, adding an extra (vertical) dimension to the excitement of watching. Watching becomes a form of dancing when people sway these lampposts violently from side to side in time to the music, themselves attracting quite a lot of attention.

In these examples, simulative performances staged in public spaces draw audiences. The practical desire to see better prompts the new use of arches, lampposts and bollards. It is not the specific form of these objects which inspires engagement; rather, the teenagers 'make do' with whatever street furniture is around. Playful competition and vertigo on these pieces of street furniture is a consequence of a new conception of how these items can be made more functional, and the difficulties of achieving that function. The audience members' attempts to use the arch and the lamppost for a new function subsequently act as a stimulus to other desires, which then give rise to further playful engagement with these props. Their play arises as a specific, spontaneous realization and enactment of a combination of desires: to see a performance, to outdo each other, to be up high, to test their strength and poise, to show themselves off.

Bollards, ostensibly dull but tremendously robust elements which are found in many public spaces, also support other forms of play. A teenager exiting the Melbourne Exhibition Centre on Southbank leapfrogs a bollard about 1200mm high, letting out an exclamation of delight. Further along Southbank Promenade, two young brothers leapfrog their way along the bollards which mark the edge of the paving. The design of these particular bollards, 800mm high, incorporates a light housed in a wide flat-topped cylinder, which is not particularly conducive to the pivoting hand action of leapfrogging. By contrast, the teenager trying to watch the performance in the Bourke Street Mall attempts to balance himself on a slippery dome-topped bollard. In both cases, the desired form of play is at odds with the planned form of the prop. Play does not only happen where physical conditions best support it; play takes place so long as human bodies are capable of harnessing the environment to explore their desires and satisfy their goals.

The diversity of stimuli and human desires collected together in urban public spaces means that any piece of street furniture which is designed to serve a particular need is likely to become creatively redeployed in the service of a range of different, novel desires. The potential of these kinds of props is activated by the diversity of people, their abilities and needs, and by the events which stimulate them. Playful uses of street furniture such as benches, bike arches and bollards depend heavily upon their locational context. These items are replicated throughout the central city. As the standard props of urban space, they can be put to use on a wide variety of stages which frame different kinds of exposure to other actors and audiences, and different challenges and risks. Street furniture provides the most exciting, the most exploratory, and the most risky examples of play with props. The difficulty and danger of these unplanned uses is also the thrill; it provides something to test oneself against. Play with street furniture is particularly transgressive, because it acts out a dialectical critique of the notion that the city, and all of its parts, are designed to 'function', to optimize practical needs.

The function of objects

Benjamin noted that people often encounter objects in the city in a state of distraction (Savage 1995). In urban space, objects are removed from familiar contexts and brought together in strange new juxtapositions. Urbanism brings objects up close, heightening people's sensory experience of their formal properties. Urban conditions can both mythify and demythify the meanings of objects which surround us, stimulating both fantasy and memory (Gilloch 1996). Through these transformations, objects become freed from their status as instruments for rational function.

Objects in the urban landscape prompt creative, exploratory engagement in at least four ways. Public artworks enable escapes from rational behavior because they lack 'function' in the strict sense; they do not help achieve any specific practical ends. Objects which are designed for playful activities can engender more kinds of engagements than their designers had in mind. Street furniture, which is designed around particular, limited functions, has formal properties which also provide affordances for a much wider range of actions. People's playful actions often appear to pursue the enjoyment of distinctive environmental properties as an end in itself. In addition, the publicness of the setting of an object also promotes its utilization for play. The props which get played with the most in Melbourne are those located on the city's three main pedestrian routes. They can all easily be viewed from a distance, moved towards and touched.

Exposure to props in public spaces can awaken and serve a wide range of human desires. Lefebvre suggests that in the city,

> objects must answer to certain 'needs' generally misunderstood, to certain despised and moreover 'transfunctional' functions: the 'need' for social life and a centre, the need and the function of play, the symbolic function of space.
>
> (Lefebvre 1996: 195)

Observations of play using various fixed objects in public settings reveal a rich scope of interrelations between perception, imagination, bodily action and spatial form. Furthermore, the varied playful, exploratory actions of people around props lead to the discovery of new possible interrelations. Play activities with props show people serving their desires for sensory pleasure, escape into imagination, testing bodily limits and engaging with strangers. These small and intricate pieces of built form are concentrated examples of the rich potentials for sensation and movement which exist throughout the urban landscape. The diverse acts of play with props which explore their varied materials, textures and shapes suggest that physical properties are a significant stimulus to non-instrumental behavior. Such properties increase in importance when people's engagements with props are not constrained by instrumental concerns such as comfort and function.

Playful actions using public artworks and street furniture lend these props a new sense of usefulness. Competitive social games organized around the fountain on Southbank and on top of benches, bicycle racks and bollards show how play action temporarily lends new, non-instrumental significance to these props: they often become game pieces to be captured from opponents. These playful applications exist in a tension with the programmed and preconceived functions and meanings of these objects. Play involving *Architectural Fragment* and two anthropomorphic artworks also establishes a tension between intended playful representations and potential playful functions of props. All these props frame the possibility of dialectical confrontations between function and pleasure, risk and safety, production and waste, comfort and bodily tension, possibility and restraint, received and produced meanings. The casino fountain and *The Three Businessmen* illustrate a number of ways that the design of props can actively promote playful exploration. But design can never resolve these tensions: that is for each individual to address, through their own playful experience. People will always play with props in creative and unexpected ways which challenge and expand their usefulness and meaning. The active role of the user in defining the possibilities of a prop and the responses of an ever-changing audience make play more compelling and rewarding.

Conclusion

Fun follows form, fun follows 'function'

Play means many different things to different people in various situations, and its potential is continually changing in space and over time. Play is not a fixed phenomenon. It cannot easily be reduced to a specific set of experiences. It takes its shape according to circumstances, and the physical setting is among those circumstances.

This book has concentrated on playful forms of activity in urban public spaces as a way to better understand the relations between the design of the built environment, the special social conditions which characterize the city, and people's perceptions and behavior. Playful actions show that urban public spaces provide new experiences and produce new social relations which are often non-instrumental, active, unexpected and risky. These ludic potentials are significant among the reasons that people gather together in urban public spaces. People can of course act playfully merely because it is their will, and neither social etiquette nor physical constraints can easily stand in their way. But play is always a fun relation to circumstances, not an outright struggle against them or a complete disregard of them. The various kinds of play illustrated here show how the practical, sensible, conventional, regular forms and functions of urban spaces provide 'useful' germination points for a great diversity of play.

Play is a product of possibility, but it is also a driver. Playful acts are within that part of the social *oeuvre* where people step beyond instrumentality, compulsion, convention, safety and predictability to pursue new and uncertain prospects. Society's aspirations are broad, and playful uses of urban spaces highlight just how much human–environment relations are enhanced by people's flexibility, imagination, skill and effort and by the support of other users around them. Play emphasizes the diversity, tension and contradiction among social practices, their circumstantiality and creativity, all of which contribute to the dialectical development of the social whole.

The idea of play as dialectical action links the minutiae of actual material practices to a much broader critique of cities and culture which argues that social life continues to evolve through dialectical negotiation with emergent forms of simulation, exploitation and discipline (Buck-Morss 1991; Debord

1994; Lefebvre 1991b, 1996). Playful acts often pay little heed to the instrumental concerns which urban designers typically aim to serve, such as comfort, stability and legibility; play flouts the subordinating, rational order, fixity and determinism which characterize urban design practice. But acts of play often engage difficulties and complexities of social space rather than ignoring them. For example, people confront the challenge of efforts to prevent them climbing or rolling over streets, furniture and buildings, rather than simply stopping or moving away.

Looking at play both theoretically and through observation of individual behaviors demonstrates that desire, needs, personal growth and freedom are not just abstract ideals confined to the realm of ideology, or instrumental goals over which actors strategize. Rather they are social experiences which arise and are constituted through practices. They are felt through the playing body. Perceptions and interactions in the public spaces of the city inspire and give shape to desire, dialectically.

Play illustrates a quite different set of relations between perceptions, intentions, actions and objects which encourage a reconsideration of both the aims and the linearity of the design process. Usually defined needs are fed into a design program which is used to generate a finished design that has an acceptable range of uses, and a management regime enforces that scope by imposing controls. But the 'playability' of urban space highlights the continual contingency and fragility of all designed, determined solutions. This view of public space is intention-centered rather than object-centered; it begins from the supposition that people have diverse desires for experience but are not necessarily or exclusively driven by particular desires or needs. When someone encounters a setting in an open, playful state of mind, they may for example inductively explore the physical potential of the space, generating bodily sensations; they may perform a use that they perceive, taking advantage of potentials which the space offers for a desired outcome, for example communicating with an audience; or they may respond in unexpected ways to conditions such as crowding, slipperiness or unfamiliarity. The process of discovering and developing playful use of public spaces is a dynamic one, and for this reason public space design should seek to also be incremental and open-ended, flexible and reworkable.

This conclusion seeks to open up a variety of ways in which the planning, design and management of the urban public realm might be kept more 'at play'. It does so by exploring three main themes of environment–behavior relations: function (ways a space does and does not support uses); publicness (free access to space); and performance (opportunities to act in front of others and express meanings). These themes provide a critical reflection on the theoretical framework of play mapped out in Chapter 2.

A fourth, overarching theme of discussion is the ways in which the elements of urban structure and their organization as urban space give shape to playful possibility. The playful potential of certain urban spatial arrangements has clear links back to a range of older and more practical theory about how cities function at the micro-scale, specifically the work of Lynch, Jacobs, Alexander and Gehl. Play provides a grounds for re-evaluating well-worn notions of good city form which are linked to efficient wayfinding, economic vitality, safety, comfort, and notions of 'primary uses' and 'necessary activities' (Jacobs 1961; Gehl 1987).

These four themes present an invitation and a challenge to policy-makers, planners and designers to adjust the values, objectives and design and management strategies through which they organize urban space. These themes are posed in terms of dialectical oppositions between conventional and more playful definitions of the aims of design of urban space. These oppositions do not outline inherently good or bad objectives. Rather, the themes suggest where, how and why those responsible for creating and managing spaces should perhaps open up and explore possibilities rather than seeking always to control, produce or prevent particular outcomes.

Making public space more useful

Observation of play illuminates the sheer diversity of behavior in public spaces, well beyond any clear, simple and precise definitions of function or efficiency. The ways in which people experience the environment surrounding them are not merely instrumental; they are often exploratory, whimsical, unsystematic and wasteful of energy. The wealth of play's contribution to people's everyday experience brings into question the idea of defining too narrowly the productive 'function' of the built environment; play is also a function. One area for policy action is for planners and designers to consider and expand upon the widest range of uses which a space might possibly have.

To allow for play, spaces need to be a little bit luxurious, in the sense of accommodating actions beyond a strict program. A public open space is, by its nature, 'an incomplete space, one that is endlessly "completed" by the people who use it'; the finding of new uses, and the process of compromise through which they are accommodated, enhance what public space means (CABE Space and CABE Education 2004: 13). Unprogrammed spaces, ill-defined spaces and loose elements are of course risky propositions with no guaranteed benefits. Yet such responses recognize that public space is in its very conception a risk, a luxurious provision of space which may or may not prove to be a cost-effective investment.

Luxurious space for play very often arises as an unintended consequence of economic spatial planning and narrow, rigidly monofunctional design. This emphasizes the dialectical production of play and the limits of design as a determinant of behavior. Leftover, underdesigned, difficult and abandoned spaces such as alleyways, floodways, tunnels, the undersides of bridges, the backs of billboards, the edges of footpaths, kerbs, excessive plazas, locked doorways, staircases, blank wall surfaces, buildings' loading docks and undercrofts provide many of the best opportunities for play precisely because they often do not have a function and their affordances are unknown (Carr et al. 1992; Stevens and Dovey 2004; Franck and Stevens 2006). These are in effect uneconomical, luxurious spaces. Investments have been made in shaping them, but they are outside the functional, managed environment; serious uses and meanings are absent. Sometimes the official 'writing off' and neglect of spaces is only very temporary, such as a closed doorway during the night or a street when the red signal holds back traffic. But the spontaneity, creativity and continual transformation of play emphasize that allowing space for play does not necessitate permanent and fixed provisions. In fact, the greatest luxury is the prospect that spaces will continue to be renewed, released, broken up or deserted, inspiring new actions. The ramshackle, irregular or inaccessible nature of these sites (such as the top of the Southbank bridge in Figure 4.4) is part of the challenge.

Lynch and Carr (1995b: 425) suggest that open space policy requires 'criteria that go beyond optimizing economy of use'. Some playful uses of the public realm are valuable to users even though they sometimes consume a lot of time, space, money or other resources:

> Simply because new activities disturb custom or are similar to real dangers, they often seem dangerous without being so . . . if we encourage a developmental world, we must exercise greater social control and also be able to restrain that control, pending the appearance of real dangers. There will be protests and reactions.
>
> (Lynch and Carr 1995b: 428)

However, spaces should be accommodating to diverse action, and designers should not go to expensive, truly wasteful efforts to suppress particular, perfectly legal activities. For example, if steps and handrails in public places are made robustly skateable, and not merely strong enough for the pressure of the hand, they will last. The same applies to making street furniture sittable and climbable.

This is not the same thing as suggesting that spaces should be wasteful, because loose, underdetermined spaces serve a more robust and inclusive notion of instrumentality and functionality. They support a much wider range of actions than spaces which are designed for a very limited range of functions and which fit those functions very closely. Luxurious spaces are actually often better utilized, because they support so many different activities for so many different people. While 'redundant' urban layouts and 'super-fluous landscapes' (Nielsen 2002) are in one sense not optimally utilized, they also provide spaces which remain available for unknown future uses. The values of spaces and activities to users are not predetermined, but are themselves constantly being produced – if spaces allow for new possibilities to emerge.

The observations of play in this book show that space designed to satisfy a particular, narrowly defined range of functions and to frame tidy, predictable social relations almost always has unintended, non-instrumental consequences for perception and behavior, and very often turns out to be conducive to a broad range of other, more playful 'functions'; 'the city . . . having been reduced to the status of a device . . . has no meaning but . . . as place of free enjoyment' (Lefebvre 1996: 126). People are incredibly perceptive and inventive in putting space to a variety of uses in play. The smooth floor of Block Arcade facilitates the unanticipated gliding of the family on in-line skates (Figure 4.11). Flag-wavers harness the wind (Figure 6.17). Some people use artwork as a ramp (Figure 8.5); others see a fenced-off, precipitous landing above the river as an ideal spot for a barbecue (Figure 6.19). Skaters 'make use of' spatial features which were never conceived in terms of non-instrumental, risky actions. The playful 'use value' of space is far broader than function (Lefebvre 1991b).

What makes play a particularly interesting basis for analyzing public space use is that play activities cut across the grain of instrumental, expected uses of space, they run against common sense and everyday 'good form' (Simmel 1950). Use is not necessarily efficient, practical or rational. It embraces oppositions between risk and safety, production and waste, comfort and bodily tension. Spaces provide challenges for those who want to seek them out and confront them. Use implies user definition of function through action, user choice and user control. Action defines the possibilities of function, and not just the other way around.

All these examples are supplementary to intended functions, suggesting that design might specifically seek to enhance the 'secondary' 'functionality' of public spaces and their furnishings. In particular, transgressive and risky behaviors suggest that policy may need to counteract the disamenity of many current design controls and design solutions which exclude particular actors or prevent particular desired actions. Reducing public leisure activity to defined functions reduces what it means to be at leisure, in part because leisure is a domain of individual choice and control (Lefebvre 1991a, 1991b).

In terms of location, non-instrumental uses of public space often lie closely adjacent to very functional uses: on the margins, just above or beyond or behind everyday 'useful' settings. One great benefit of placing public art in the midst of public space is that perceptual and bodily distractions can readily occur in the context of everyday functional activity. Street furniture also gets played on because it is on stage. Play as vertigo thrives at the boundaries and margins between spaces. Play is fluid, continuous, adaptive. It engages with sudden shifts and discontinuities within space and learns to master them. Thresholds and intersections are critical zones where a space suited to one function spills into another space which has different possibilities. At the intersection or threshold itself, people have to stop or slow down; there is a distortion in the functional time and space rhythms of the city which changes potential for action. These settings also show the inherent spatial conflicts among the functional needs of various directions and modes of travel through public space. The bike lift during Critical Mass (Figure 4.10) shows that the functionality of the intersection space is only provisional, and is dependent upon obedience to functional conventions.

Caillois' (1961) four basic types of play help to spell out the diversity of things people are actually doing in public spaces. Play in the forms of competition and vertigo highlights that people have close, multi-sensory experiences of the material urban fabric, and sometimes choose to test their body against physical limits. These bodily experiences of space involve taking risks, rather than maximizing certainty and comfort. Perceptions of space and actions in space are not just 'banal, practical [and] effective' (Lynch 1981: 131–32), but also affective. Rather than treating the structure of the built environment as a tool, play pursues the sensory perception of distinctive environmental properties as a pleasurable end in itself.

Making public space more open

Huizinga (1970) suggests that play required a place apart from the everyday. However, public play occurs in public settings and involves members of the public. Playful uses of space suggest the benefits of a more generous notion of the publicness of public space, a wider, more democratic considera-tion of users and potential users. The public is a complex, changing and always controversial concept, but the design of public spaces seldom comes to terms with its inherent tensions. A public space which is well ordered, predictable, safe and comfortable feels that way only to a relatively small and uniform group of users; and even for them, not all the time. Spaces are public in different ways and to different extents. For example, the entry steps of a building are not really public. Even many fully public spaces often have discreet forms of management which assign those spaces discrete and carefully modulated functions, such as the part of the street right-of-way between the kerbs.

A space is truly public only if it can be accessed and used freely by all people. This is what Lefebvre (1996) refers to as 'The Right to the City'. Vital public space requires attention to the kind of public which might and can be drawn there, and to concentrate on new ways to attract users. Open space is ' "open" not because it is free of buildings and covered with plants, but in the sense that it is uncommitted to prescribed users' (Lynch and Carr 1995b: 424). Open space has to be 'open to the freely chosen and sponta-neous actions of people . . . it has no necessary relation to ownership, size, type of use, or landscape character' (Lynch 1995: 396). Such definitions emphasize the absence of regulation over people, their freedom and oppor-tunity to come and play in spaces. A diverse public is necessary to develop the scope of use as fully as possible.

The intensity, variety, pleasure and opportunity that people find through various types of play suggest that less resources should be devoted to design interventions and management strategies which increase disamenity, controls and rules and reduce the scope of users. Even if sometimes these forms of control turn out to be dialectically productive of new and interesting forms of play, there are clearly already more than enough difficult, ugly spaces in cities. Evidence of people playing on stairs and window ledges, in alleyways and on sculptures shows that they are already aware of these opportunities. The challenge is to create playful spaces which provide a similar scope of opportunities, in a non-deterministic way that allows people to discover and establish their own forms of action, and to create functional spaces which can better accommodate physical, exploratory and imaginative and social forms of play.

The need to ensure physical access and availability does not imply that spaces should be endlessly open. Too much openness deprives people of control and affordances (Gibson 1979; Lynch 1981). Making spaces accommodating to a wide public requires ensuring that one group's activities do not inhibit other groups, and this means spaces generally need to provide 'articulations that allow mixed occupancy and use' (Lynch and Carr 1995a: 415). Ideally, most public spaces should have more edges within them, more distinct zones and more variation in character, and more contents. Play in public focuses attention on the great variation of ways that people like to position themselves in relation to edge conditions of urban space, at many different scales, in order to regulate their level of exposure to other people and to other stimuli. To ensure adequate levels of stimulation, people often seek the tension of boundaries within public space which are 'defined and yet not too defined' (Alexander et al. 1977: 349). Edges and thresholds are popular sites for play precisely because of their looseness and variability of experience. They provide something to work with and work against. Public space is often too open, too even and too safe.

Public space policy might also change by expanding the physical and temporal horizons of spaces which is considered to be public and in use. The five spatial contexts for play examined in Chapters 4 to 8 illustrate that urban space provides playful possibilities at a range of scales, from the overall urban structure down to the smallest details of steps and statues. To enhance prospects for play requires focusing on the qualities and affordances of space, and not necessarily on assigning and designing discrete spaces. Although parks and plazas vary in size, the range of types of planned open space is very small. It is 'a fact of life that most play does not take place on sites formally designated as play spaces' (Department for Culture, Media and Sport 2004: 10). There are a great many spaces used for leisure functions 'which are not green on planner's maps'. Available space for play 'has no necessary relation to ownership, size, type of use, or landscape character', and can include all 'the negative [i.e. unbuilt], extensive, loose, uncommitted' space in the city; although it cannot include many 'green' and 'open' spaces which are private, programmed or inaccessible (Lynch 1995: 396–97).

A great number and variety of play activities occur around building thresholds, which are not always thought of as spaces in their own right, and which are not truly public. Play broadens the scope of public space in a literal sense, by bringing attention to physical margins and in-between places. Because play has a dialectical relation to other everyday activities and experiences, spaces where people play are often leftover or superfluous spaces located right next to spaces with fixed and delimited functions; heightening the contrast of playful use and seriousness, and allowing their interaction, as when performers stand near busy footpaths and intersections (Figure 6.2). Open spaces which can be used playfully include those which have low levels of investment, regulation and surveillance and those which are underdesigned. Policy could aim to ensure the continued existence of spaces which are 'slack', underutilized and available for informal uses. The potential usefulness of many spaces only becomes apparent through people's dynamic playful appropriation.

Observations of playful activities pursuing vertigo foreground those physical spaces, or at least parts of spaces, which are not easy and obvious to 'use'. People climb on things that are difficult to climb on, follow paths which are treacherous, and squeeze between objects. Skateboarding is one activity that has a very complex and particular set of geometries (Borden 2001b; Woolley and Johns 2001). Many skateboarders use spaces other than purpose-built skate parks because the design of such facilities offers too limited a range of opportunities for action. These acts all suggest the need for more exploration of prospective leisure sites, by policy-makers and designers as well as by users.

The many playful activities which are characterized by movement along pathways through the city (parades, races, explorations on cycles and skates) highlight that functional access routes can themselves become sites of play. Pathways, especially minor, indirect and hidden pathways, make 'other' settings more accessible, creating networks of small and varied spaces which expand the range of experiential and behavioral opportunities (Lynch 1995). Functional pathways can form the basis for sequences of leisure activities, turning a necessary activity into a desirable one and bringing people at play into contact with a great many other people. Some of this potential is illustrated by cycleways, historic trails, tourist highways and fitness circuits in parks. Former canals and rail corridors provide the opportunity to transform a lost function into new experiential possibilities, as is currently being undertaken for New York's abandoned freight railway viaduct, the High Line (Ryan 2006).

Planning policy can also expand upon the temporal horizon within which public space use is considered. A significant and distinctive range of play activities occur at night, on weekends and at other times when the primary designed use of sites is in abeyance, for example skaters using the thresholds of office buildings.

Leisure means the passing of time free from compulsion, and in particular from the need to engage in productive activities (Goodale and Godbey 1988). Sometimes people play by remaining stationary in settings which are actually designed for constant movement, blocking doorways, closing streets to traffic or sitting on steps. In other cases, people start to play because a setting prevents them from continuing their instrumental activities, even if only for a short time, such as when pedestrians wait at traffic signals. These are among the 'specific moments in an individual's life when he is most open to new possibilities. These are not task-oriented periods, but times of leisure, holidays, commuting, waiting' (Lynch and Carr 1995b: 428). Thus play can be enhanced through management of urban time as well as space. Designers should recognize that the uses of a space often continue and change after the intended users have gone home.

A great many non-functional playful activities run against conventional time-use patterns of public settings. Policy and design for the public realm might easily be adjusted to better serve late and early uses, adding secondary 'functionality' to spaces. Many play activities exist in the temporal margins because they conflict with prime, normative functions, but different design or management might allow them to take place concurrently. Policy might also seek to better provide for unusually fleeting uses of public spaces. Some users obtain considerable enjoyment from moving very quickly through a place. Sometimes playful use of a space begins from someone pausing momentarily in mid-step when they are distracted. A more comprehensive understanding of temporal planning might also consider that design is 'never finished' (Francis 2003: 66), that new needs and new uses are always evolving, and so design policy needs to consider how to best provide for the possibility of ongoing alterations to public open spaces and their programs.

Making public space more public

In the city, people are always on display to strangers. Playful performance is an important function of public space, one which produces pleasure, skills and identities. Like other functions, seeing and being seen is largely dependent on spatial affordances. Physical space contributes to the framing of roles in public drama, defining where and who is 'on stage'. Certain stages are obvious, because changes in height enhance visibility, barriers prevent intrusion and backdrops block out distractions. But in other cases, public performances take shape through the ambiguities and difficulties of spatial definition and control. There are a tremendous range of different sites where people perform: some are spontaneous and others are not; some interfere with functional use of space and others do not. Various acts of play illustrate constant tensions between displays which are instrumental and those which are not, between playful performances and more instrumental uses of public space, and between the actions of various parties who each enjoy the freedom of public space.

Performances in busy spaces like train stations and street intersections often come into direct conflict with highly instrumental needs. Yet steps, thresholds, intersections and the edges of paths can all easily be appropriated for performances because their functional users are so transient. The children who pose as mannequins in the shop window (Figure 6.8) illustrate how a solid physical barrier between actors and their intended audience can shape the nature of a performance. The barrier reduces communication only to visibility; there is no touch, no movement, no sound. It is precisely these conditions which the play gets shaped around. This act transforms the potential of a window in public view from instrumental commercial display to non-instrumental performance. Spaces such as this which are designed only to be representational remain among the most open to playful practices. In a diametrical situation, the man who dances in front of the Sanity CD store (Figure 7.9) relies on the seepage of an instrumental musical performance across a threshold to lend atmosphere to his own display. While spectacular images in the city are framed for distant, passive perception, this music is engaged intensely through the body, in the presence of others. The adjacency of this man's performance to a public audience also permits him to draw them into the act.

Simulations rely on audiences; competitions depend on competitors. Not all public performances are solo, and some arise through interactions among strangers. Rather than requiring separation, these performances occur on stages where people are brought together in public. The city's potential to inspire play rests in part on the diverse ways it brings active playful performances together on stage with the everyday lives of strangers. Gradations of physical visibility and accessibility, and behavioral freedom and social involvement help determine how and how much people are drawn away from their predetermined actions into new, unfamiliar, non-instrumental performances.

People like the tourists at London's Leicester Square and Melbourne's Bourke Street Mall (Figures 6.9 and 6.10), who step from the audience to dance out on the stage, transgress the nominal, indefensible boundaries of the stage. Instrumental performances can stimulate other playful acts which elude instrumentality. Intersections and thresholds are two types of conspicuous public locations where strangers have close, unplanned bodily encounters in the presence of onlookers which can turn dramatic.

From the perspective of the audience, people are often attracted to places where they can sit or stand comfortably to watch the activities of others (Gehl 1987). A significant corollary documented in this book is that performative acts of play occur where people are present to witness them. Looking at play as performance underscores the significant dialectical interaction between active and passive uses of urban public space.

Audiences matter, whether they have chosen to watch or not, whether they are pleased, thrilled, annoyed or confronted. The responses of onlookers add to the meaning, and thus the desirability, of play, partly because the public context is so dynamic and unpredictable. Public drama develops dialectically, emerging from an interplay between spatial opportunities, the desires of various participants, and behaviors which mobilize these potentials.

It is clear from observing simulative play that representation is among the functions of public spaces, meaning the symbolism of sites themselves, but also sites' capacities for people to develop and produce meanings: what Lefebvre (1991b) calls 'representational spaces'. The many playful public performances which occur in urban space highlight that people constantly create new meanings for themselves and others around them, contest existing meanings, or ignore meanings in favor of the physical opportunities which representational objects present. Meanings are just as important to the play of audiences as they are to performers. 'Perception is a creative act' (Lynch 1981: 131) and audiences respond. The figural and contextual contributions which urban designers make clearly cue certain possibilities for representation, but urban design is different from other arts because it gets inhabited by the viewer and put to functional, interpretive and expressive uses: 'This is the space which the imagination seeks to change and appropriate' (Lefebvre 1991b: 39). Ultimately the potentials which a built form has for representation are determined when they are brought together with the desires, ideas and skills of actors, physical opportunities for communicating, and reception by audiences. Meanings are produced dialectically out of the interrelations of these factors.

Public parades emphasize that streets are not just for efficient circulation; they frequently have representational functions which need to be considered when making functional modifications to those streets. Although the meanings of a street can be imposed through regular practice, they are not always agreeably received by disparate audiences, and can be contested and rewritten. The protest to bring back the Birdman Rally (Figure 4.13) reframes Melbourne's street carnival by acting within the space and the time which claims to represent carnivalesque, playful freedom.

Another way in which practices engage spatially embodied meanings is when people's play defines new representational spaces. Wedding photographs taken in the city (Figures 7.6 to 7.8) illustrate how exploratory, unconventional, playful behavior can produce the built environment as a symbolic landscape. The wedding couple who climb up into a window ledge (Figure 7.8) emphasize people's creativity in reinterpreting urban form. In wedding photos, people are not just acknowledging but actually inscribing meanings within the built environment. People's interests in producing unanticipated meanings, often in unexpected places, suggests there are both limits and difficulties in designers' frequent tendency to thematize spaces, to emphasize the public's role as a passive audience, and to prevent playful augmentations of meaning.

The feel of the city: urban structure and phenomenology

In addition to increasing the usefulness, openness and publicness of individual settings, it is also important to consider how play can be promoted by the general organization of the urban structure: the arrangement of spaces and activities and the ways people move among them. Observations show that urban play does not only happen in playgrounds; it is widely distributed.

The places where people go and where they play depend on their perceptions of the city. Lynch (1960) notes that when people move through cities, they make use of mental maps of the urban structure. Their comprehension of urban spatial structure and their ability to use it develop over time through the accumulation of their perceptions. Lynch found that people's mental images of cities are composed of five fundamental, reasonably invariant elements: *paths, edges, districts, nodes* and *landmarks*. People remember these elements because they can be clearly perceived and they have a strong relation to people's actions. Some of the kinds of spaces where playful behavior tends to occur are similar to those foregrounded by Lynch; others are very different. A comparison of these two studies of users' experience of urban structure enables a more comprehensive and robust understanding of the phenomenological and behavioral significance of urban structure, particularly because play frames a far broader domain of urban spatial experience.

For designers seeking to understand how urban spatial morphology frames perception and behavior, Lynch's methodology has several significant limitations. He assesses a person's perceptions about urban form only in relation to two narrow, predetermined functional objectives: correctly identifying their relative location, and moving toward their goal. He does not ask people about their perception of spatial conditions which relate to their other activities. Most importantly, he is not interested in the spatiality of social interactions. In mapping people's paths through the city, Lynch draws only on people's memories. He overlooks the bodily act of walking itself, which is clearly integral to the experience of space, and ignores the impact which the presence and actions of numerous other pedestrians in a 'swarming mass' has on each individual's perception and behavior in public (de Certeau 1993).

Lynch was studying the structure of images of cities at a large scale, as remembered and drawn, and not the structure of spaces as inhabited and used. By focusing on formal identity, he understates the fact that urban form's symbolic meaning and its role in the actions of the observer greatly heighten its significance. Caillois' (1961) typology of play shifts attention to the more physical, athletic forms of play, rather than receptive, more passive experiences of the urban scene. Lynch's drawn and mental maps privilege visibility, and in particular the detached, abstracted distance of the aerial viewpoint (de Certeau 1993). Urban dwellers do not only use vision; they need 'to hear, to touch, to taste and . . . to gather these perceptions in a "world"' (Lefebvre 1996: 147).

Lynch was merely analyzing the structure of how people remembered cities, but many scholars and designers take Lynch's work for normative prescription (Sternberg 2000). His findings, intended to shed light on the 'problem' of becoming lost, have become a justification and a template for producing visually well-ordered spaces which promote a well-ordered society (Debord 1994). Wayfinding is a functional objective which has limited importance to the general well-being of society, and legibility is not an unassailable end in itself (Carmona et al. 2003). Lynch himself noted of congruence, transparency and legibility in urban settings that 'none of these . . . are absolute desiderata' (Lynch 1981: 143). He observes that

> the function of a good urban environment may not be simply to facilitate routine trips, not to support meanings and feelings already possessed. Quite as important may be its role as a guide and a stimulus for new exploration.
>
> (Lynch 1960: 109–10)

He later called attention to 'our delight . . . in ambiguity, mystery . . . surprise and disorder' (Lynch 1991: 250–52). Surprises are possible in public spaces because so many strangers are moving and acting independently there. But relatively little is known about the various ways in which urban structure frames more impractical aspects of urban experience: the unexpected, unfamiliar and incomprehensible, spontaneity, distraction and risk.

Three kinds of spatial elements appear to be important both for practical cognition in wayfinding and for the diverse, unplanned activities of play: paths, nodes/intersections and edges/boundaries. The significance of these three kinds of spaces for perception, memory and playful action indicates their key importance in physical, perceptual and psychological terms. They are the fundamental topological structure of space in relation to movement and visibility, defining (respectively) the continuity of space, choices of direction, and enclosure (Norberg-Schulz 1971, 1980).

Play reveals that the ways people feel urban structure are somewhat different from the ways they see and remember it. Lynch's two other elements, districts and landmarks, are conceptual 'gestalt' components of an individual's large-scale mental map which have a very limited role in spatial experiences other than the practical task of wayfinding from memory. Two elements of urban spatial topology which are important to play are lacking from Lynch's image of the city: props and thresholds. These are concrete spatial elements experienced up close with the body and which come between various occupants of the urban space, mediating their interactions.

People do see districts as actual physical settings, to the extent that they are places 'which the observer mentally enters "inside of"' (Lynch 1960: 47). Districts have a sense of interiority. The phenomenology of entering into a district, or any recognized social setting, necessarily involves the spatial experience of crossing a threshold. The large amount and variety of playful behavior seen to occur around thresholds is testament to the diverse experiential possibilities they frame. People often linger and make the most of experiences which are available on thresholds. The thresholds of busy public buildings stimulate playful displays and sometimes playful confrontations.

Thresholds with lower levels of use offer people the opportunity to control their level of exposure. In contrast to Lynch's focus on practical, goal-directed perceptions of the city, the long, open facades of video game arcades frame the distraction of many passers-by. Publicly accessible thresholds seldom neatly separate the realms of inside and outside. Sounds, smells and warmth drift back and forth across these permeable boundaries, washing over passing bodies, providing sensory information. Unlike images, these other kinds of sensations are not often spatially fixed, they continually circulate, intermingle and change their structure. People often encounter them very incidentally. People's involvements with the complex conditions around thresholds indicate that urban spatial cognition is less regular, less well-managed and more playful than Lynch's study of images suggests.

Props may easily be overlooked as a part of the environmental structure because they are small. The inclusion of such localized 'urban detail' as signs, trees, doorknobs and distinctive window curtains within Lynch's category of landmarks highlights the importance of small-scale objects in facilitating urban wayfinding (Lynch 1960; De Jonge 1962). While landmarks can provide 'trigger cues whenever turning decisions must be made' as people navigate through the city (Lynch 1960: 83), close-up, multi-sensory experience of the materiality of space can also trigger many other kinds of memories and emotions and desires, and thus stimulate a wide variety of human actions.

Props can have representational contents which catalyze simulative play. In London's Leicester Square, the street performers who dress up and stand on pedestals pretending to be statues (Figure 7.3) are in part stimulated by the numerous memorial statues located there, including Shakespeare, Hogarth, Newton and Chaplin. Play at Melbourne's busiest pedestrian intersection, Bourke Street Mall and Swanston Walk, also focuses around the statues located there: *The Three Businessmen Who Brought their Own Lunch*. At both sites, 'elements located at junctions . . . automatically . . . derive special prominence from their location' (Lynch 1960: 73); but these elements do much more than just enhance wayfinding. Lynch's consideration of how the city image's various elements combine in people's experience is very brief, and designers need to be aware of how the interrelation between various urban spatial elements can create complex sets of opportunities.

The fact that the prescribed meanings of public artworks are often ignored in favor of bodily experiences which their physical form enables is particularly significant in relation to Lynch's study: interest in immediate, tactile experience dominates the visual register, memory and instrumentality. Lynch mentions that kinaesthesia can aid memorability, for example when people move along a curved or sloping path. Skateboarders illustrate a far stronger link between spatial form, kinaesthesia and action, as their freely chosen movements produce pleasurable bodily sensations. Skaters, among others, explore minor, everyday functional urban design features such as steps, handrails, planter boxes, bollards, bicycle racks and benches, seeing them as challenging physical landscapes which can be explored with the body.

These various observations of playful bodily engagement with objects located in urban space provide a preliminary sketch of the scope of behavior which can arise through people's active perception of urban landscapes. These small and intricate pieces of built form are concentrated examples of the rich potentials for sensation and movement which exist throughout the urban landscape. In contrast to Lynch's depiction of landmarks as distant images which provide a dominating, fixed structure to the experience of moving through urban space, different props may over time capture attention, lose their significance or change their function depending on how they are encountered and put to use. The purpose of a prop is performed by a variety of actors in front of changing audiences. The design of objects which are placed in urban spaces needs to recognize this continual expansion of their functionality; including accepting that not all users of public art will notice or respect its representational properties.

Further research is needed to explore in more detail the microgeography of urban space, the way in which built elements structure human experience and movement within the scale of the body's reach, the behavioral importance of particular properties such as slope, texture and temperature, perhaps by focusing on specific non-instrumental behaviors. Such research could draw upon a wealth of earlier work which examines the social and perceptual dimensions of body space (see Chapter 3) as well as a range of recent work focusing on the special nature of perception in urban space (e.g. Rodaway 1994; Savage 1995; Latham 1999; Borden 2001b).

Props and thresholds highlight the closeness, richness and dynamism of bodily experience of urban space, and the spatial framing of roles for other people who are involved in one's actions. Observations of playful behavior involving props and thresholds reveal a rich scope of interrelations between perception, memory, intention, symbolism, human bodies, actions and spatial form. Furthermore, the varied playful, exploratory actions of people using these elements lead to the discovery of new possible interrelations. Rather than perceiving objects and spaces merely as instrumental means to a practical end, people's playful actions around thresholds and props serve desires for sensory pleasure, escape into imagination, testing bodily limits, and engaging with strangers.

Play in public space highlights a range of person–environment relationships quite different from Lynch's but rather complementary: purposeful, efficient movements have their counterpart in those which are whimsical, unsystematic and wasteful of energy; actors' clear and decisive responses to unique spatial forms contrast to those which are equivocal, vague and changing. Looking at play illuminates the very wide scope of the 'affordances' which the urban built environment offers for human perception and also for action, both practical and otherwise (Gibson 1979).

Density, mixing and disorder

Play cannot always be predicted and cannot always be designed for. Many of the observations of playful activity in this book highlight the spontaneous and unexpected nature of people's experiences and actions. Lynch (1960) saw the image of the city as quite static and invariant, whereas Benjamin depicts the city as being labyrinthine, shifting and dialectical, and therefore much more conducive to play (Buck-Morss 1991; Gilloch 1996). It is in play as chance that urban structure can most clearly be seen to shape experience. Urban design has a significant role in promoting the possibility of chance, discovery and diversity and risk.

Play appears to be facilitated by many of the same properties of urban structure which urban designers believe to help sustain and enhance the general vitality and robust diversity of social practice. Neighborhoods built at small scale encourage people to walk. Urban environments which are composed of highly permeable, densely interconnected circulation routes – 'ringy' spaces – provide more opportunities to change direction when moving through the city and allow spaces and people to be encountered in different sequences, undermining the possibility of strict control over movement, increasing unpredictability and bringing strangers into contact more frequently (Jacobs 1961; Dovey 1999). Frequent cross-streets foster 'incubation, experimentation and many small or special enterprises' (Jacobs 1961: 183); this is true for businesses because it is also true at the level of individual actions. The spontaneity and diversity of perceptions and possibilities which are presented at cross-streets promote playfulness in people's responses to them.

Not only a mix of primary uses but also buildings of different ages are essential for the economic vitality of a neighborhood (Jacobs 1961). Multiple attractors are likely to bring people to an area at different times, and because people walking to a particular destination are likely to be distracted along the way, such localized 'pools of use' help sustain smaller, economically marginal activities which depend on walk-in business. Under the right circumstances, the spatial and temporal distribution of people's 'necessary' activities within a neighborhood – their rational, efficient 'functions' – can also support a very wide range of other 'optional', impractical activities such as play (Gehl 1987). Planning thus needs to prevent monofunctional and class-specific buildings such as hotels, churches, universities, offices and cultural centers from monopolizing and constraining the vitality of open spaces which are only potentially public (Stevens 2006). Conversely, planning should aim to cluster facilities which will draw diverse social groups together around public spaces, so that people are more likely to engage strangers when they are coming and going.

The close spatial relations between play and serious activities, and the fact that play often occurs in very functional places such as doorways, highlight that all publicly accessible outdoor spaces need to be thought of as integral parts of the mix of use areas within a neighborhood. Designated open spaces serve as primary destinations for only a relatively small number of users and a relatively narrow range of activities (Jacobs 1961; Lynch 1995). Incidental playful uses of public settings are, by contrast, numerous and varied. Providing a variety of underprogrammed, underdesigned, accommodating spaces in the midst of major activity areas such as transport interchanges, places of work, shopping centers and cultural venues can promote a much wider variety of playful behavior than providing 'free' space in isolation.

Buildings of different sizes and ages in a neighborhood ensure varying rents and diverse layouts. Continual piecemeal redevelopment of the urban fabric guarantees that there are always spaces available for the emergence of new economic activities with varying financial resources and varying, somewhat unpredictable spatial needs (Jacobs 1961). The same argument holds true for outdoor spaces and non-economic activities: play relies on varied spaces. Because they are diverse, idiosyncratic and creative, play activities seek out spatial opportunities beyond the limits of economy. Optional activities of leisure are more sensitive and responsive to spatial qualities than work and reproductive activities are. Without the abstract, deferred benefit of economic gain, people tend to relax and play only when and where environmental conditions suit their desires (Gehl 1987).

Jacobs (1961) argues it is the combination of concentration, fine-grained urban structure, mixing of uses and heterogeneous spatial development which makes a neighborhood vital and robust. An urban neighborhood with these qualities is vital because it does not have a tight, singular relation between form and function. It is loose and adaptable, and somewhat luxurious. It remains open to non-practical action. It does not suffer when one key function moves away or is not in operation. But more than this, the complexity of a neighborhood can also help to promote new and diverse forms of social behavior. The larger vitality or 'playability' of urban life lies in the complex opportunities created by its spaces. Rather than careful, precise management of hierarchies of primary and secondary functions, robust urban form opens up many potential interrelations between functions, spaces and non-functional activities.

Alexander (1965) suggests that the functionality of an urban environment actually lies in its provision for many potential functions, the creation of a large number of possible relations between human actions and physical settings. He explains that various elements of the physical environment become combined together in different ways within a myriad of different patterns of human action. The relation between forms, actions and the needs which actions serve is many-to-many; there are a great number of possible permutations, forming a 'lattice'. By contrast, 'tree-like' thinking which correlates individual functions with specific spaces or built forms in a very instrumental, positivist and reductive fashion kills off the diversity and robustness of urban life.

Alexander's (1965) own example of this more indeterminate functionality is a street corner with a newspaper stand and a traffic light. The red light momentarily stops a pedestrian from fulfilling one function; the newspaper stand makes possible another 'function', albeit one which is only a latent desire in the mind of some passers-by. Gehl (1987) describes similar shifts in terms of necessary activities giving way to optional ones. Needs give way to desires. Gehl emphasizes that environmental conditions have to be right for optional activities to occur; the shift in motivation is because of a possibility which is provided by the built environment. Non-instrumental, playful actions highlight that people's desires to act are dynamic because they are at least in part produced by the opportunity and stimulus of the spatial and social context. It is thus inappropriate to think of the environment as necessarily being designed to best serve predefined functions. A more general vitality of urban life lies in the many potential interrelations between spaces and an infinite variety of freely chosen activities which may be improvisatory and spontaneous and not just instrumentally rational. The behavioral open-endedness of Alexander's (1965) idea of an urban spatial 'lattice' is well illustrated by Jacobs' (1961) description of the 'place ballet' on the sidewalks of Greenwich Village:

> Under the seeming disorder of the old city . . . is a marvelous order for maintaining . . . the freedom of the city . . . This order is all composed of movement and change . . . an intricate ballet . . . The ballet of a good city sidewalk never repeats itself from place to place, and in any one place is always replete with new improvisations.
>
> (Jacobs 1961: 50)

A pedestrian paused on the street corner by the newspaper stand and the traffic light might enter into conversation or change her mind about where she is going. The presence of other objects in the setting, such as public artworks or stairs, can provide further potentials for action, as can other people who are moving in other directions with other intentions. To understand and optimize the richness of the place ballet requires attention to the stage details: to props, to entries and exits, and to the immense variety of acts which discover and engage these spatial opportunities. The functionality of open space cannot be predetermined as a 'program' or prevented. Functionality is, rather, latent in a space's formal properties, planned or otherwise, and is constantly discovered and utilized by everyday actors who are both 'rich in needs' and in means to pursue them (Caillois 1961; Lefebvre 1991b: 165). Any program for open space can be only a preliminary script for the continuously unfolding drama of open space use. Settings provide opportunities for people to discover and invent new patterns of use which might suit their varying needs.

Classic urban design theory confirms that urban settings which are less rigidly ordered and more complex in structure make it harder for designers and managers to predict and control the activities of users. Urban density, diversity and complexity allow playful possibilities, and also stimulate them. Another way in which urban design can open up playful behavioral opportunities is through improved provision for vertigo. Rather than always striving to make public spaces comfortable, harmonious, relaxing and easy to use, consideration might be given to designing some aspects of urban layouts and individual spaces such that they are difficult for users themselves to comprehend and control. Vertiginous environments which challenge people's bodily limits include places which are large, steep or high, such as bridges, ledges, balconies and stairs, many of which can be climbed and jumped from, and can also include those which are smooth and slippery and allow people to move fast or in unfamiliar directions. Feelings of vertigo can also be prompted by environments which engage and stretch people's perceptions, whether sight, balance, orientation, noise, temperature or smell.

There are potentially great benefits in allowing public spaces to be more arresting to the senses, more disorderly and more risky, but these uncertain, varied, impractical and personal benefits are seldom weighed against the quantifiable financial costs of injuries and conflicts in the use of public space. The prevalence of play as vertigo shows that the users of public spaces often feel that 'calculated' risks are 'worth taking'. For encounters with risks and difficulties to be playful means that actors remain able to perceive and choose whether to engage with them or not. Apparent risks and dangers are actually good safety features: people know they need to concentrate if they are to manage risks that are large and obvious. Presenting risks in a playful context gives people the chance to explore, to test limits, and to thereby develop greater competence. It is often when people carelessly assume that spaces are absolutely safe and easy to negotiate that accidents occur.

Play as vertigo also embraces a range of actions which other users of spaces see as disorderly or even find obnoxious. Enhancing the scope of playful use of public open spaces might also include responding positively and proactively to the observable fact that some users of public spaces sometimes want to be physically confined, make noise, dance wildly and even damage things. Rather than always attempting to prevent or displace such behavior, high-performance, optimally inclusive urban design could strive to accommodate vertigo as part of the reality of everyday life, helping to ensure such actions remain harmless, contained and playful.

The Situationist International (SI) provide one means for developing a better understanding of the disorderly, spontaneous, unfamiliar, incomprehensible, distracted and risky nature of urban experience, through their research in the field of psychogeography (see Chapter 1). In sharp contrast to the functionalist aims, methods and theories pioneered by Lynch, the SI's approach emphasizes that cities are experienced in time and in motion by active subjects with bodies, histories and complex feelings, and not by purely rational consciousness. The Situationists' own experiences with *dérive* highlight aspects of a city's physical organization which attract attention for reasons which are not clearly functional; sites and routes which are boring or disappointing, forgotten or bypassed; those where detours from one's goal are necessary and where unexpected and unintended events happen. The SI's interpretation is that urban space, rather than being logical, has 'radical discontinuities and divisions', locations which are mundane, useless and unknown, all of which suggest the potential for playful appropriation (McDonough 1994: 69). They note that spaces that change frequently and those that are disorienting make routine, customary behavior and inhabitation impossible and made imaginative play more likely. They also note that social needs and desires can be served by particular lighting levels, color, acoustics, odor, texture, temperature and moisture; and that these environmental conditions can also induce emotional states and thereby stimulate activities.

The Situationists' critique zoning, 'the concentration camp organization of life', where space is segregated for work, residence and consumption and according to household type and social class (Author unknown 1996a: 118). They argue that such structures serve a very reductive notion of 'function', making urban life structured and repetitive, reducing free time and increasing alienation. Urban planning and design generally tend to organize and circumscribe human action in space, to limit the risk of incursions upon intended, conventional urban behaviors by making alternative acts more difficult. However, functional, orderly urban space does not mean that play disappears. Practices of play do not arise through a tidy, rational chain linking intentions, actions, functions and outcomes; they do not necessarily have a fixed purpose, and they are thus difficult to regulate through rational strategies. Attempts to circumscribe kinds of play which are risky, messy, or controversial through design are easily eluded. Being independent of instrumental needs and outside rational calculation of purpose, being diverse and creative in their implementation, practices of play are not vulnerable to the kinds of rules, fears and expectations which cause people to regulate their serious practices. Players such as street performers, wedding couples and skaters are in continuous motion through the city, scouring it for spatial opportunities to express playful needs.

The dialectics of play

The inherent appeal of playful activity in public spaces, as both an experience and an expression of freedom, human potential, rich and immediate sensation, exploration and self-realization, presents an invitation to designers and planners, on both a professional and personal level, to create environments which might better serve the desire to play. But looking at playful behavior, both theoretically and from observation, with its spontaneity, fickleness, willfulness, disorderliness, difficulty and danger, and its sheer variety of forms and aims, also complexifies and problematizes the practicability of designing for play, if design is thought of as matching form to function.

The analyses of playful uses of space in this book are not intended to suggest tidy, practical solutions for either promoting or curtailing play through spatial design. The support which urban space provides to play is always broader than any conscious intention; play is excessive. While intended to fulfill needs sensibly and rationally, urban design amenities also inevitably enhance prospects for the pursuit of desire, disorder, destruction and immediate pleasure. The benefits of public space and the diverse, playful 'uses' of public space cannot be predetermined; they are defined through action, just as desires themselves can arise through action. Urban public space generates possibilities, it stimulates actions and reactions. It enables the discovery of new potentials and new needs. It is in such ways and within such places that city life reveals its promise of freedom.

Playful use shows that urban space is not neatly staged, purely functional or unambiguous in meaning. But neither is social life itself teleological; new goals and means are constantly being revealed. One of the main fallacies of instrumentalism is the presumption that spatial forms serve functions which are immanent, sovereign and fixed, and that behavior is always a consequence of prior, reflective thought. In fact behavior is in very significant ways responsive to, and generated by, the stimulus of physical and social context. Rather than thinking of play as discrete forms of experience which have specific characteristics and which the built environment can support, or perhaps even cause, a more useful insight is to think of playfulness as an ingredient of people's general experience of urban space, as one of the aspects and moods of all life in cities; an ever-present potential within the way places are and the way people feel and act.

This book foregrounds play's potency as an internal and developmental critique of everyday life (Lefebvre 1991a). Playful behaviors illustrate the internal tensions between any given action and what people like to think of as being 'normal' uses and users of public space. Urban space as lived is a product of the contradictions which arise between instrumental function, social reproduction, and the diverse scope of other interests and desires that comprise everyday urban life. People who interrupt parades (Figure 4.13), upstage street musicians (Figure 6.10), turn police barricades into display racks (Figure 6.14), block doorways (Figure 7.2) and put cigarettes in statues' mouths (Figure 8.2) show that tension itself is part of the delight of urban experience, as well as being a powerful generator of new possibilities for action. Many new meanings and forms of practice are born of contradiction. Functional, manipulative design often prompts play dialectically, producing desires for other kinds of actions, because physical and social confrontations are often intrinsically enjoyable. Rules and conventions are adapted, revised and made up as people play; 'urban life tends to turn against themselves the messages, orders and constraints coming from above. It attempts to appropriate time and space by foiling dominations, by diverting them from their goal' (Lefebvre 1996: 117). These 'disruptive' playful behaviors reveal and expand potentials for social and spatial experience.

Play often evolves as a form of resistance to environmental determinism, both behavioral and semantic. Signs with regulations prohibiting certain activities make it starkly apparent that transgressions of behavioral norms either occur or are suspected; that serious, rational activities and play are in spatial tension. These are signs of possibility: anywhere authorities have to place a sign prohibiting an activity, the public can be sure that it is both possible and popular to undertake the activity there. Behavioral restrictions built into spaces, such as ridges added to a bridge to prevent people climbing it (Figure 4.5), metal lugs attached to steps and ledges to inhibit skateboarding, and large sculptures introduced to obstruct performances (Figure 6.7) also pinpoint patterns of use. Such changes do not often thwart action because they do not thwart the desire to act. They tend, rather, to transform or displace play. The more complex the boundaries, rules and obstacles within the environment, and the more the environment is transformed, the more terrain there is for users to explore and contest. The four forms of play, competition, simulation, chance and vertigo, showcase the variety of ways that people foil the dominations which contemporary urban society has stamped onto space. Various acts of play show how people can accommodate the contradictions of possibility and restraint.

Playful behavior highlights people's conscious interest in the affordances and the limitations of the physical and symbolic landscape, of their own body and mental powers, and of their relations to strangers in the city. To say that play is a developmental critique of everyday life is to emphasize that play does not only critique life in the abstract, it produces new experience: play is the actualization of freedom, adventure, creativity and discovery, although not in any fixed sense. Play is a 'performative experiment', where the continuing, cumulative actions of individuals in space continuously define social potential by acting it out using the expressive, affective and perceptual powers of the body (Thrift 1997). Urban public space is the key ground for this playful production of the different and the new. This is part of why people are attracted to cities; it is part of what public spaces are for. Competition and simulation, which both rely on audiences, illustrate the necessity of ongoing dialogue between presentations and their affirmation by audiences of strangers. Tensions among playful uses such as watching and participating are themselves very productive, generating new possibilities.

The observations in this book focus largely on the more active forms of play. This reflects an interest in how people act in relation to their physical environment and each other, and in people's proactive efforts to establish or adjust those relations. The focus on action has been sharpened by employing Caillois' (1961) formulation of competition, simulation, chance and vertigo. People can be a passive audience to all these kinds of play events, yet they emphasize that someone is doing something in the city in order to shape their own experience. The active slant also does justice to the fact that cities are especially active places. The dialectical condition of urban public space is necessarily linked to bodily action and the regulation of bodies.

This book highlights several particular kinds of urban settings where desires for play tend to become manifest in bodily practices. Paths, intersections and thresholds tend to be sites of play in motion and many competitive, simulative and unexpected encounters between strangers. Paths also emphasize the relationships between sequences of experiences in time, space and memory. Boundaries and props tend more often to be places for stopping, and often involve close bodily engagements with material space. Boundaries and thresholds make use of the differentiations of space to structure playful social relations. But this is not an exhaustive list of settings for play, and there is no fixed correspondence between built form, human motivation and social behavior. Values, ideas, actions and spaces define one another and are constantly produced through each other. The design of urban public spaces enters into an extraordinarily complex matrix of perceptions, attitudes, meanings and behaviors. The city is not fixed, not an ideal diagram, but a system of potential in flux. Built form is the most stable part of this system, but this just makes its design and management the most problematic. The power which a setting provides to relate to other people, to signify or to act is complex and never guaranteed.

To see urban life as dialectical means to view the functional, representational and sociological impact of urban design as being not only far-reaching but also imprecise and ambiguous. The ever-present, responsive possibilities of play suggest that the absence of a clear, rational program for urban design should not necessarily be considered a problem to be solved; indeterminacy, incompleteness, looseness and risk can also present opportunities. Urban design should be loose, because in cities, behavior and meanings are slippery, they remain at play. In truly public spaces, there will always be vagaries, flexibilities and conflicts; all have their merits. For cities to be vital, urban design needs to recognize the unfunctional and the fleeting, the partial and the uncertain; and to be provocative and invite exploration, by admitting overlap, exposure, doubt and risk.

References

Alexander, C. (1965) 'The City is Not a Tree', *Architectural Forum*, 122(1): 58–62 and 122(2): 58–61.

Alexander, C., Ishikawa, S., Silverstein, M., Jacobson, M., Fiksdahl-King, I. and Angel, S. (1977) *A Pattern Language: Towns, Buildings, Construction*, New York: Oxford University Press.

Arendt, H. (1958) *The Human Condition*, Chicago, IL: University of Chicago Press.

Author unknown (1996a) 'Critique of Urbanism' in L. Andreotti and X. Costa (eds) *Theory of the Dérive and Other Situationist Writings on the City*, Barcelona: Museu d'Art Contemporani de Barcelona ACTAR.

Author unknown (1996b) 'Unitary Urbanism at the End of the 1950s' in L. Andreotti and X. Costa (eds) *Theory of the Dérive and Other Situationist Writings on the City*, Barcelona: Museu d'Art Contemporani de Barcelona ACTAR.

Bakhtin, M. (1984) *Rabelais and his World*, Bloomington, IN: Indiana University Press.

Ball, E. (1987) 'The Great Sideshow of the Situationist International', *Yale French Studies*, 73: 21–37.

Bataille, G. (1985) 'The Notion of Expenditure' in *Visions of Excess*, Minneapolis, MN: University of Minnesota Press.

—— (1988) *The Accursed Share, Vol. 1*, New York: Zone Books.

Bateson, G. (1987) 'A Theory of Play and Fantasy' in *Steps to an Ecology of Mind*, Northvale, NJ: Aronson.

Bauman, Z. (1993) *Postmodern Ethics*, Oxford: Blackwell.

Benedikt, M. (1979) 'To Take Hold of Space: Isovists and Isovist Fields', *Environment and Planning B*, 6: 54–59.

Benjamin, W. (1974) *Gesammelte Schriften*, Frankfurt am Main: Suhrkamp Verlag.

—— (1997) 'Paris, Capital of the Nineteenth Century' in N. Leach (ed.) *Rethinking Architecture: A Reader in Critical Theory*, London: Routledge.

—— (2006) *Berlin Childhood around 1900*, Cambridge, MA: Belknap Press of Harvard University Press.

Borden, I. (1998) 'Body Architecture: Skateboarding and the Creation of Super-Architectural Space' in J. Hill (ed.) *Occupying Architecture*, London: Routledge.

—— (2000) 'Skateboarding and Henri Lefebvre: Experiencing Architecture and the City' in *Habitus 2000: A Sense of Place – Conference Proceedings*, Perth: Curtin University of Technology.

—— (2001a) 'Another Pavement, Another Beach: Skateboarding and the Performative Critique of Architecture' in I. Borden, J. Kerr and J. Rendell (eds) *The Unknown City: Contesting Architecture and Social Space*, Cambridge, MA: MIT Press.

—— (2001b) *Skateboarding, Space and the City: Architecture and the Body*, Oxford: Berg.

Bourdieu, P. (1977) *Outline of a Theory of Practice*, Cambridge: Cambridge University Press.

—— (1984) *Distinction: A Social Critique of the Judgment of Taste*, Cambridge, MA: Harvard University Press.

—— (1990) *The Logic of Practice*, Stanford, CA: Stanford University Press.

—— (1998) *Practical Reason: On the Theory of Action*, Cambridge: Polity Press.

—— (2000) Lecture to Habitus 2000 Conference, Perth, Australia, 7 July.

Brown-May, A. (1998) *Melbourne Street Life*, Melbourne: Australian Scholarly.

Buck-Morss, S. (1991) *The Dialectics of Seeing: Walter Benjamin and the Arcades Project*, Cambridge, MA: MIT Press.

CABE Space and CABE Education (2004). *Involving Young People in the Design and Care of Urban Spaces: What Would You Do with This Space?* London: Commission for Architecture and the Built Environment.

Caillois, R. (1961) *Man, Play and Games*, New York: Free Press of Glencoe.

Campo, D. (2002) 'Brooklyn's Vernacular Waterfront', *Journal of Urban Design*, 7: 171–99.

Carmona, M., Heath, T., Oc, T. and Tiesdell, S. (2003) *Public Spaces – Urban Places: The Dimensions of Urban Design*, London: Architectural Press.

Carr, S., Francis, M., Rivlin, L. and Stone, A. (1992) *Public Space*, Cambridge: Cambridge University Press.

Castle, T. (1986) *Masquerade and Civilization*, Palo Alto, CA: Stanford University Press.

Cavan, S. (1966) *Liquor License: An Ethnography of Bar Behavior*, Chicago, IL: Aldine.

City of Melbourne (1985) *Streets for People: A Pedestrian Strategy for the Central Activities District of Melbourne*, Melbourne: City of Melbourne.

—— (1997) *Grids and Greenery Case Studies*, Melbourne: City of Melbourne.

City of Melbourne and Gehl Architects (2004) *Places for People*, Melbourne: City of Melbourne.

Clark, K. and Holquist, M. (1984) *Mikhail Bakhtin*, Cambridge, MA: Harvard University Press.

Cohen, S. and Taylor, L. (1978) *Escape Attempts: The Theory and Practice of Resistance to Everyday Life*, Harmondsworth: Pelican.

Constant (Niewenhuis) (1997/1959) 'A Different City for a Different Life', *October*, 79: 109–12.

Crawford, M. (1992) 'The World in a Shopping Mall' in M. Sorkin (ed.) *Variations on a Theme Park: The New American City and the Design of Public Space*, New York: Hill and Wang.

Dargan, A. and Zeitlin, S. (1990) *City Play*, New Brunswick, NJ: Rutgers University Press.

Darnton, R. (1984) *The Great Cat Massacre and Other Episodes in French Cultural History*, New York: Basic Books.

Debord, G. (1994) *The Society of the Spectacle*, New York: Zone Books.

—— (1996a) 'Introduction to a Critique of Urban Geography' in L. Andreotti and X. Costa (eds) *Theory of the Dérive and Other Situationist Writings on the City*, Barcelona: Museu d'Art Contemporani de Barcelona ACTAR.

—— (1996b) 'Situationist Theses on Traffic' in L. Andreotti and X. Costa (eds) *Theory of the Dérive and Other Situationist Writings on the City*, Barcelona: Museu d'Art Contemporani de Barcelona ACTAR.

—— (1996c) 'Theory of the Dérive' in L. Andreotti and X. Costa (eds) *Theory of the Dérive and Other Situationist Writings on the City*, Barcelona: Museu d'Art Contemporani de Barcelona ACTAR.

De Certeau, M. (1984) *The Practice of Everyday Life*, Berkeley, CA: University of California Press.

—— (1993) 'Walking in the City' in S. During (ed.) *The Cultural Studies Reader*, London: Routledge.

De Jonge, D. (1962) 'Images of Urban Areas: Their Structure and Psychological Foundations', *Journal of the American Institute of Planners*, 28: 266–76.

—— (1967) 'Applied Hodology', *Landscape*, 17: 10–11.

Department for Culture, Media and Sport (2004) *Getting Serious about Play: A Review of Children's Play*, London: Department for Culture, Media and Sport.

Dovey, K. (1999) *Framing Places: Mediating Power in Built Form*, London: Routledge.

During, S. (1993) 'Editor's Introduction (Michel de Certeau: "Walking in the City")' in *The Cultural Studies Reader*, London: Routledge.

Eckstut, S. (1986) 'Solving Complex Urban Design Problems' in A. R. Fitzgerald (ed.) *Waterfront Planning and Development*, New York: American Society of Civil Engineers.

Edensor, T. (1998) 'The Culture of the Indian Street' in N. Fyfe (ed.) *Images of the Street: Planning, Identity and Control in Public Space*, New York: Routledge.

Fast, J. (1971) *Body Language*, London: Pan.

Foucault, M. (1977) *Discipline and Punish: The Birth of the Prison*, London: Penguin.

—— (1997) 'Space, Knowledge and Power – Interview with Paul Rabinow' in N. Leach (ed.) *Rethinking Architecture: A Reader in Critical Theory*, London: Routledge.

Francis, M. (2003) *Urban Open Space: Designing for User Needs*, Washington, DC: Island Press.

Franck, K. and Stevens, Q. (2006) 'Tying Down Loose Space' in K. Franck and Q. Stevens (eds) *Loose Space: Possibility and Diversity in Urban Life*, London: Routledge.

Gehl, J. (1987) *Life between Buildings: Using Public Space*, New York: Van Nostrand Reinhold.

Gehl, J. and City of Melbourne (1994) *Places for People*, Melbourne: City of Melbourne.

Gehl, J. and Gemzoe, L. (1996) *Public Spaces, Public Life*, Copenhagen: Danish Architectural Press.

Gibson, J. (1979) *The Ecological Approach to Visual Perception*, Boston, MA: Houghton Mifflin.

Giddens, A. (1979) *Central Problems in Social Theory: Action, Structure and Contradiction in Social Analysis*, London: Macmillan.

Gilloch, G. (1996) *Myth and Metropolis: Walter Benjamin and the City*, Cambridge: Polity Press.

Goffman, E. (1959) *The Presentation of Self in Everyday Life*, New York: Anchor.
—— (1971) *Relations in Public: Microstudies of the Public Order*, New York: Basic Books.
—— (1972) 'Fun in Games' in *Encounters: Two Studies in the Sociology of Interaction*. London: Penguin.
—— (1980) *Behavior in Public Places*, Westport, CT: Greenwood.
—— (1982) 'Where the Action Is' in E. Goffman (ed.) *Interaction Ritual*, New York: Pantheon.
Goodale, T. and Godbey, G. (1988) *The Evolution of Leisure: Historical and Philosophical Perspectives*, State College, PA: Venture.
Gottdiener, M. (1985) *The Social Production of Urban Space*, Austin, TX: University of Texas Press.
—— (1997) *The Theming of America: Dreams, Visions and Commercial Spaces*, Boulder, CO: Westview.
Habermas, J. (1997) 'Modern and Postmodern Architecture' in N. Leach (ed.) *Rethinking Architecture: A Reader in Critical Theory*, London: Routledge.
Hall, E. T. (1966) *The Hidden Dimension: Man's Use of Space in Public and Private*, London: Bodley Head.
—— (1973) *The Silent Language*, New York: Anchor.
Hannigan, J. (1998) *Fantasy City: Pleasure and Profit in the Postmodern Metropolis*, New York: Routledge.
Harvey, D. (1989) *The Urban Experience*, Oxford: Blackwell.
Hillier, B. and Hanson, J. (1984) *The Social Logic of Space*, New York: Cambridge University Press.
Huizinga, J. (1970) *Homo Ludens: A Study of the Play Element in Culture*, London: Temple Smith.
Jackson, P. (1988) 'Street Life: The Politics of Carnival', *Environment and Planning D*, 6: 213–27.
Jacobs, J. (1961) *The Death and Life of Great American Cities*, New York: Vintage.
Jinnai, H. (1995) *Tokyo: A Spatial Anthropology*, Berkeley, CA: University of California Press.
Judd, D. and Fainstein, S. (eds) (1999) *The Tourist City*, New Haven, CT: Yale University Press.
Kearns, G. and Philo, C. (eds) (1993) *Selling Places*, Oxford: Pergamon.
Kofman, E. and Lebas, E. (1996) 'Lost in Transposition' in H. Lefebvre, *Writings on Cities*, Oxford: Blackwell.
Kostof, S. (1992) *The City Assembled: The Elements of Urban Form throughout History*, London: Thames and Hudson.
Kotanyi, A. and Vaneigem, R. (1996) 'Elementary Program of the Bureau of Unitary Urbanism' in L. Andreotti and X. Costa (eds) *Theory of the Dérive and Other Situationist Writings on the City*, Barcelona: Museu d'Art Contemporani de Barcelona ACTAR.
Lancy, D. and Tindall, B. A. (eds) (1977) *The Study of Play: Problems and Prospects*, West Point, NY: Leisure Press.
Latham, A. (1999) 'The Power of Distraction, Tactility and Habit in the work of Walter Benjamin', *Environment and Planning D*, 17: 451–73.
Lefebvre, H. (1971) *Everyday Life in the Modern World*, New York: Harper and Row.

—— (1987) 'An Interview with Henri Lefebvre', *Environment and Planning D*, 5(1): 27–38.

—— (1991a) *Critique of Everyday Life, Vol. 1*, 2nd edn, London: Verso.

—— (1991b) *The Production of Space*, trans. D. Nicholson-Smith, Oxford: Blackwell.

—— (1996) *Writings on Cities*, trans. and eds E. Kofman and E. Lebas, Oxford: Blackwell.

—— (with Kristin Ross) (1997a) 'Lefebvre on the Situationists: An Interview', *October*, 79: 69–83.

—— (1997b) 'The Everyday and Everydayness' in S. Harris and D. Berke (eds) *Architecture of the Everyday*, New York: Princeton Architectural Press.

Lennard, S. and Lennard, H. (1984) *Public Life in Urban Places*, Southampton, NY: Gondolier.

—— (1995) *Liveable Cities Observed*, Carmel, CA: Gondolier.

The Lettrist International (1996) 'Skyscrapers by the Roots' in L. Andreotti and X. Costa (eds) *Theory of the Dérive and Other Situationist Writings on the City*, Barcelona: Museu d'Art Contemporani de Barcelona ACTAR.

Lofland, L. (1998) *The Public Realm: Exploring the City's Quintessential Social Territory*, New York: Aldine de Gruyter.

Lutfiyya, M. N. (1987) *The Social Construction of Context through Play*, Lanham, MD: University Press of America.

Lyman, S. and Scott, M. (1975) *The Drama of Social Reality*, New York: Oxford University Press.

Lynch, K. (1960) *The Image of the City*, Cambridge, MA: MIT Press.

—— (1977) *Growing Up in Cities: Studies of the Spatial Environment of Adolescence in Cracow, Melbourne, Mexico City, Salta, Toluca, and Warszawa*, Cambridge, MA: MIT Press.

—— (1981) *Good City Form*, Cambridge, MA: MIT Press

—— (1991) 'Reconsidering the Image of the City' in T. Banerjee and M. Southworth (eds) *City Sense and City Design: Writings and Projects of Kevin Lynch*, Cambridge, MA: MIT Press.

—— (1995/1965) 'The Openness of Open Space' in T. Banerjee and M. Southworth (eds) *City Sense and City Design: Writings and Projects of Kevin Lynch*, Cambridge, MA: MIT Press.

Lynch, K. and Carr, S. (1995a/1979) 'Open Space: Freedom and Control' in T. Banerjee and M. Southworth (eds) *City Sense and City Design: Writings and Projects of Kevin Lynch*, Cambridge, MA: MIT Press.

—— (1995b/1968) 'Where Learning Happens' in T. Banerjee and M. Southworth (eds) *City Sense and City Design: Writings and Projects of Kevin Lynch*, Cambridge, MA: MIT Press.

Macarthur, J. (1999) 'Tactile Simulations: Architecture and the Image of the Public at Brisbane's Kodak Beach' in R. Barcan and I. Buchanan (eds) *Imagining Australian Space: Cultural Studies and Spatial Inquiry*, Nedlands, WA: University of Western Australia Press.

McDonough, T. (1994) 'Situationist Space', *October*, 67: 58–77.

Markus, T. (1993) *Buildings and Power: Freedom and Control in the Origin of Modern Building Types*, London: Routledge.

Martins, M. (1982) 'The Theory of Social Space in the Work of Henri Lefebvre' in R. Forrest, J. Henderson, and P. Williams (eds) *Urban Political Economy and Social Theory: Critical Essays in Urban Studies*, Aldershot, UK: Gower.

Maslow, A. (1943) 'A Theory of Human Motivation', *Psychological Review*, 50: 370–96.

Mean, M. and Tims, C. (2005) *People Make Places: Growing the Public Life of Cities*, London: Demos.

Mouledoux, E. (1977) 'Theoretical Considerations and a Method for the Study of Play' in D. Lancy and B. A. Tindall (eds) *The Study of Play: Problems and Prospects*, West Point, NY: Leisure Press.

Mumford, L. (1961) *The City in History*, New York: Harcourt Brace.

—— (1996/1937) 'What is a City?' in R. LeGates and F. Stout (eds) *The City Reader*, London: Routledge.

Nasaw, D. (1985) *Children of the City*, New York: Doubleday.

Nielsen, T. (2002) 'The Return of the Excessive: Superfluous Landscapes', *Space and Culture*, 5: 53–62.

Nietzsche, F. (1968) *The Will to Power*, New York: Vintage.

Norberg-Schulz, C. (1971) *Existence, Space and Architecture*, New York: Praeger.

—— (1980) *Genius Loci: Towards a Phenomenology of Architecture*, London: Academy Editions.

Percy, W. (1975) 'The Loss of the Creature' in *The Message in the Bottle*, New York: Farrar, Straus and Giroux.

Phillips, E. B. and LeGates, R. (1981) *City Lights: An Introduction to Urban Studies*, New York: Oxford University Press.

Plant, S. (1992) *The Most Radical Gesture: The Situationist International in a Postmodern Age*, London: Routledge.

Radley, A. (1993) 'The Elusory Body and Social Constructionist Theory', *Body and Society*, 1(2): 3–23.

Rodaway, P. (1994) *Sensuous Geographies*, London: Routledge.

Rojek, C. (1995) *Decentering Leisure: Rethinking Leisure Theory*, London: Sage.

Ryan, Z. (ed.) (2006) *The Good Life: New Public Spaces for Recreation*, New York: Van Alen Institute.

Sadler, S. (1998) *The Situationist City*, Cambridge, MA: MIT Press.

Sandercock, L. and Dovey, K. (2002) 'Pleasure, Politics and the Public Interest: Melbourne's Waterfront Revitalization', *Journal of the American Planning Association*, 68: 151–64.

Savage, M. (1995) 'Walter Benjamin's Urban Thought', *Environment and Planning D*, 13: 201–16.

Scheflen, A. (1972) *Body Language and Social Order*, Englewood Cliffs, NJ: Prentice-Hall.

—— (1976) *Human Territories: How We Behave in Space-Time*, Englewood Cliffs, NJ: Prentice-Hall.

Sennett, R. (1971) *The Uses of Disorder: Personal Identity and City Life*, Harmondsworth: Penguin.

—— (1974) *The Fall of Public Man*, Cambridge: Cambridge University Press.

—— (1994) *Flesh and Stone: The Body and the City in Western Civilization*, New York: Norton.

Shields, R. (1991) *Places on the Margin: Alternative Geographies of Modernity*, London: Routledge.

Simmel, G. (1950/1917) 'Sociability (An Example of Pure, or Formal, Sociology)' in *The Sociology of Georg Simmel*, Glencoe, IL: Free Press.

—— (1997/1903) 'The Metropolis and Mental Life' in N. Leach (ed.) *Rethinking Architecture: A Reader in Cultural Theory*, London: Routledge.

Sommer, R. (1969) *Personal Space: The Behavioral Basis of Design*, Englewood Cliffs, NJ: Prentice-Hall.

Sorkin, M. (ed.) (1992) *Variations on a Theme Park: The New American City and the Design of Public Space*, New York: Hill and Wang.

Spariosu, M. (1989) *Dionysus Reborn: Play and the Aesthetic Dimension in Modern Philosophical and Scientific Discourse*, Ithaca, NY: Cornell University Press.

—— (1997) *The Wreath of Wild Olive: Play, Liminality and the Study of Literature*, Albany, NY: SUNY Press.

Stavrides, S. (2001) 'Spatiotemporal Thresholds and the Experience of Otherness', *Journal of Psychogeography and Urban Research*, 1(1): Online. Available: http://www.psychogeography.co.uk/spatiotemporal thresholds.htm (accessed 10 December 2001).

Sternberg, E. (2000) 'An Integrative Theory of Urban Design', *Journal of the American Planning Association*, 66: 265–78.

Stevens, P. (1991) 'Play and Liminality in Rites of Passage: From Elder to Ancestor in West Africa', *Play and Culture*, 4: 237–57.

Stevens, Q. (2006) 'The Design of Urban Waterfronts: A Critique of Two Australian "Southbanks"', *Town Planning Review*, 77(2): 173–203.

Stevens, Q. and Dovey, K. (2004) 'Appropriating the Spectacle: Play and Politics in a Leisure Landscape', *Journal of Urban Design*, 9: 351–65.

Stratford, E. (2000) 'Feral Skateboarding and the Field of Transport Planning: Some Observations on the Tasmanian Case' in *Habitus 2000: A Sense of Place – Conference Proceedings*, Perth: Curtin University of Technology.

Sutton-Smith, B. (1997) *The Ambiguity of Play*, Cambridge, MA: Harvard University Press.

Thrift, N. (1997) 'The Still Point: Resistance, Expressive Embodiment and Dance' in S. Pile and M. Keith (eds) *Geographies of Resistance*, London: Routledge.

Tuan, Y.-F. (1977) *Space and Place: The Perspective of Experience*, Minneapolis, MN: University of Minnesota Press.

Turner, V. (1982) *From Ritual to Theatre: the Human Seriousness of Play*, New York: Performing Arts Journal Publications.

Vaneigem, R. (1996) 'Commentaries against Urbanism' in L. Andreotti and X. Costa (eds) *Theory of the Dérive and Other Situationist Writings on the City*, Barcelona: Museu d'Art Contemporani de Barcelona ACTAR.

van Gennep, A. (1960) *The Rites of Passage*, London: Routledge and Kegan Paul.

Whyte, W. H. (1980) *The Social Life of Small Urban Spaces*, Washington, DC: Conservation Foundation.

—— (1988) *City: Rediscovering the Center*, New York: Doubleday.

Wirth, L. (1996/1938) 'Urbanism as a Way of Life' in R. LeGates and F. Stout (eds) *The City Reader*, London: Routledge.

Wittgenstein, L. (1958) *Philosophical Investigations*, Oxford: Blackwell.

Woolley, H. and Johns, R. (2001) 'Skateboarding: The City as a Playground', *Journal of Urban Design*, 6: 211–30.

Zukin, S. (1991) *Landscapes of Power: From Detroit to Disney World*, Berkeley, CA: University of California Press.

Index